ISO 9001:2015 in Brief

ISO 9001:2015 in Brief provides an introduction to quality management systems for students, newcomers and busy executives, with a user friendly, simplified explanation of the history, requirements and benefits of the new standard.

This short, easy-to-understand reference tool also helps organisations to quickly set up an ISO 9001:2015 compliant Quality Management System for themselves at minimal expense and without high consultancy fees.

Now in its fourth edition, *ISO 9001:2015 in Brief* consists of a number of chapters covering topics like:

- What is Quality? – An introduction to the requirements and benefits of quality, quality control and quality assurance
- What is a QMS? – The structure of a Quality Management System and associated responsibilities
- Who produces Quality Standards? – An opportunity to see how interlinked the various Standards Bodies are today
- What is ISO 9001:2015? – The background to this particular standard, how it has grown and developed over the years and what 'Annex SL' is all about
- What other standards are based on ISO 9001:2015? – Details of other standards that replicate or are broadly based on ISO 9001:2015
- What to do once your QMS is established – Process improvement tools, internal auditing and the road to ISO 9001:2015 certification

This is supported by:

- A summary of the requirements of ISO 9001:2015, including an overview of the content of the various clauses and sub-clauses, the likely documentation required and how these would affect an organization. A cross-reference to the previous ISO 9001:2008 clauses is also provided as well as a complete bibliography and glossary.

Ray Tricker (MSc, IEng, CQP-FCQI, FIET, FCMI, FIRSE) is currently working as the Senior Management Consultant for Herne European Consultancy Ltd – a company specialising in offering organisations access to a range of highly skilled and specialist consultants to help these companies enhance their business performance.

ISO 9001:2015 in Brief

Fourth edition

Ray Tricker

Routledge
Taylor & Francis Group

LONDON AND NEW YORK

First published 2001 by Butterworth-Heinemann

Reprinted 2001, 2002

Second edition 2005 by Elsevier

Third edition 2009 by Herne

Fourth edition 2016 by Routledge
2 Park Square, Milton Park, Abingdon, Oxon OX14 4RN

and by Routledge
711 Third Avenue, New York, NY 10017

Routledge is an imprint of the Taylor & Francis Group, an informa business

British Library Cataloguing in Publication Data
A catalogue record for this book is available from the British Library

Library of Congress Cataloging in Publication Data
Names: Tricker, Ray, author.
Title: ISO 9001:2015 in brief / Ray Tricker.
Description: Fourth edition. | Abingdon, Oxon ; New York, NY : Routledge, 2015. | Includes bibliographical references.
Identifiers: LCCN 2015046524| ISBN 9781138025851 (hardback) | ISBN 9781138025868 (pbk.) | ISBN 9781315774831 (ebook)
Subjects: LCSH: ISO 9001 Standard. | Quality control—Standards. | Manufactures—Quality control—Evaluation.
Classification: LCC TS156.17.I86 T75 2015 | DDC 658.4/013—dc23
LC record available at http://lccn.loc.gov/2015046524

ISBN: 978-1-138-02585-1 (hbk)
ISBN: 978-1-138-02586-8 (pbk)
ISBN: 978-1-315-77483-1 (ebk)

Typeset in Minion Pro and Optima
by Florence Production Ltd, Stoodleigh, Devon, UK

To my Grandsons
Kenneth, Joseph and James
In the hope that this book may prove
useful to them in their future chosen careers

Contents

About the author

Ray Tricker (MSc, IEng, FCQI-CQP, FIET, FCMI, FIRSE) is a Senior Consultant with over 50 years' continuous service in Quality, Safety and Environmental Management, Project Management, Communication Electronics, Railway Command, Control and Signalling systems and the development of molecular nanotechnology.

He served with the Royal Corps of Signals (for a total of 37 years), during which time he held various managerial posts culminating in being appointed as the Chief Engineer of NATO's Communication Security Agency (ACE COMSEC).

Most of Ray's work since leaving the services has centred on European Railways. He has held a number of posts with the Union International des Chemins de Fer (UIC) [e.g. Quality Manager of the European Train Control System (ETCS)] and with the European Union (EU) Commission [e.g. T500 Review Team Leader, European Rail Traffic Management System (ERTMS) Users Group Project Coordinator, HEROE Project Coordinator]; currently (as well as writing books on such diverse subjects as International Standards, Communication Electronics, Building, Wiring and Water Regulations for Taylor & Francis, Elsevier and Van Haren) he is busy assisting Small Businesses from around the world (usually on a no-cost basis) to produce their own auditable Quality and/or Integrated Management Systems to meet the requirements of ISO 9001, ISO 14001 and OHSAS 18001. He is also a United Kingdom Accreditation Service (UKAS) Assessor (for the assessment of Certification Bodies in regard to harmonisation of the Trans-European, High Speed Railway Network) and recently he was the Quality, Safety and Environmental Manager for the consultancy overseeing the multi-billion dollar Trinidad Rapid Rail System.

Currently he is working as the Senior Management Consultant for Herne European Consultancy Ltd – a company specialising in offering organisations access to a range of highly skilled and specialist consultants to help these companies enhance their business performance.

One day, he says that he might retire!!

Foreword

Last year I wrote to Ray thanking him for the work that he does on ISO 9001 (particularly for small businesses), and how his work has had a positive impact on me and the companies that I have worked for in the past 14 years as a senior managerial consultant in the electronics industry.

Now, I'm not typically forthcoming when praising authors, especially in the somewhat dry field of quality management; however, in the case of Ray's work it is well deserved. If you are lucky, you will have already stumbled across his work early in your quest to get a handle on the ISO 9001 standard and to effectively implement it in your business, whatever that may be.

I was not so lucky! Hours of trawling through the first few pages of dross that a Google search will offer up and a ream of paper later, I was little the wiser as to what ISO 9001 was all about or how I was to implement it. I thankfully resisted the temptation to outsource the problem to a consultant and compromised by buying one of Ray's ISO 9001:2000 books. That was in 2004 and I haven't looked back since, or bought another book for that matter!

Ray's common sense approach (which is not so common in my experience!) and the way he is able to précis the 'legalise' of this particular International standard into words is a very welcome and refreshing change. Ray's ISO 9001 series of books have assisted companies, both large and small, throughout the world in understanding and making use of this important quality management standard.

ISO 9001:2015 in Brief is an important book in the series and it is aimed at those people who find the contents of the ISO 9001 standard daunting, inaccessible or who just want to be told straight what's needed and wanted. It covers all of the necessary background that you will need to appreciate how useful managing quality really is, and is delivered using Ray's inimitable style. He uses humour and easy-to-understand descriptions with apt illustrations to highlight the most important aspects of this standard . . . You won't forget it and it will remain with you forever!

In short, this book is a lifesaver for anyone who is serious about implementing a quality system of life and vigour – but it is not a free lunch! Time and dedication are required to make it come alive. What Ray does is provide the ingredients and the recipe. Now it's your turn to make the meal and season to the taste of your particular business.

Bon appetit!

Jason Solloway

Preface

The ISO 9000 family is an all-encompassing series of standards that lay down requirements for incorporating the management of quality into the design, manufacture and delivery of products, services and software.

The family consists of three primary standards supported by a number of technical reports. These are:

1   **ISO 9001:2015 Quality Management Systems** – which sets out the requirements for a Quality Management System and is the only standard in the ISO 9000 family against which an organisation can be certified and registered. To achieve its main objectives, this standard requires the designers, manufacturers, suppliers and end users to possess a fully auditable Quality Management System consisting of Quality Policies, Quality Processes, Quality Procedures and Work Instructions. It is this Quality Management System that will provide the auditable proof that the requirements of ISO 9001:2015 have been and *still are* being met.

2   **ISO 9000:2015 Quality Management Systems. Fundamentals and vocabulary** – which gives an overview of the fundamentals and concepts around quality management. This standard has been extensively revised and includes all of the terms and definitions users need to understand if they plan to implement a Quality Management System to the requirements of ISO 9001:2015.

3   **ISO 9004:2009 A Quality Management approach to sustainable success of an organisation** – which focuses on how to make a Quality Management System more efficient and effective by providing guidance on achieving sustainable success through Quality Management.

 Note: please note that ISO 9004:2009 is *not* intended for certification, regulatory or contractual use.

 Author's Hint
ISO 19011:2011 Guidelines for auditing management systems – is a very useful standard which sets out guidance on internal and external audits of Quality Management Systems.

Why has ISO 9001 been revised?

In accordance with International agreements, *all* ISO standards are reviewed every five years to establish whether a revision is required in order to keep it current and relevant for the marketplace. As a consequence, the previous 2008 version of ISO 9001 has undergone a six-stage revision process in order to respond to the latest trends and be compatible with other management systems (e.g. the ISO 14001 environmental standard).

ISO 9001:2015 was eventually published in September 2015 and was immediately accepted by manufacturers, suppliers and end users as a major step forward in Quality Management.

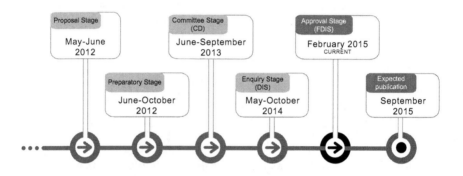

The six-stage revision process for ISO 9001:2015

One of the primary changes made to the previous edition of the standard is that as ISO 9001:2015 has been rewritten around the common framework (as detailed in Annex SL of the ISO/IEC Directives – see Ch 5.5 for detailed explanation), all new and future revisions of existing Management Systems Standards will now have the same high-level structure, identical core text, and common terms and definitions.

 Note: Although the high-level structure of Management Systems Standards cannot be changed, sub-clauses and discipline-specific text can be added.

What are the changes in this edition of *ISO 9001 in Brief* from previous versions?

During the revision process of ISO 9001, the ISO Committee (ISO/TC 176) recognised that, although it was desirable to maintain the emphasis on process management, greater emphasis would need to be placed on producing desired outputs and providing confidence in a product and/or service.

Thus a major overhaul of the existing ISO 9001:2008 was required in order to:

- provide a stable core set of requirements for the next 10 years or more;
- maintain the current focus on effective process management to produce desired outcomes;
- take account of the changes in Quality Management Systems practices and technology since the last major revision in 2008;
- make it easier to implement and use;
- provide a conformity for assessments by first, second and third parties;
- use simplified language and writing styles to ease understanding;
- provide consistent interpretations of its requirements;
- introduce *risk-based thinking*;
- place more emphasis on the seven Quality Management Principles;
- emphasise the importance of leadership;
- ensure that **everyone** is responsible for the organisation's quality.

This fourth edition in the ISO 9001 in Brief series has taken these changes into consideration and the main parts of the book are as follows:

- **What is quality?** – an introduction to the requirements and benefits of quality, Quality Control (QC) and Quality Assurance (QA).
- **What is a Quality Management System?** – the structure of a Quality Management System and associated responsibilities.
- **The history of Quality Standards** – a time map showing how Quality Control has developed over the last seven centuries.
- **Who produces Quality Standards?** – an opportunity to see how interlinked the various Standards Bodies are today.
- **What is ISO 9001:2015?** – the background to this particular standard, how it has grown and developed over the years and what 'Annex SL' is all about.
- **What other standards are based on ISO 9001:2015?** – details of other standards that replicate or are broadly based on ISO 9001:2015.
- **What to do once your QMS is established** – process improvement tools, internal auditing and the road to ISO 9001:2015 certification.

This is supported by:

- **Annex A – ISO 9001:2015 – a summary of requirements** – which is an extension of the Chapter entitled 'What is ISO 9001:2015?' and which provides a précis (in quite some detail – but don't worry, not too much!) of the actual requirements of ISO 9001:2015; an overview of the content of the various clauses and sub-clauses contained in the standard: the likely documentation that will be required and a short resumé of how these would affect an organisation. A cross-reference to the previous ISO 9001:2008 clauses are also provided.

- **Bibliography** – containing: Abbreviations and Acronyms; Reference Standards for Quality Management Systems; Glossary of terms used in Quality Management Standards; and of course, Books by the same Author!

I have also taken the opportunity to:

- copyedit and remove a large amount of cross-references to previous versions of the standard;
- include a number of semi-humorous illustrations on the assumption that '*a picture is worth a thousand words*';
- keep explanations as simple as possible in order to appeal to students, newcomers to Quality Assurance or the beleaguered executive with little time to come to terms with the subject!
- try to reduce the number of equivalent or similar terms.

For convenience (and in order to reduce the number of equivalent and/or similar terms) the following, unless otherwise stated, are considered interchangeable terms within this book:

- **product** – hardware, software, service or processed material that is the result of an organisational process that does not include activities performed at the interface between the supplier (provider) and the customer;
- **organisation** – a single person or group of people who achieve their objectives by using their own functions, responsibilities, authorities and relationships. It can be a company, corporation, enterprise, firm, partnership, charity, association or institution either privately or publicly owned. It can also be an operating unit that is part of a larger entity;
- **service** – the result of a process that includes at least one activity carried out at the interface between the supplier (provider) and the customer (e.g. the delivery of knowledge).

In the text of the book, you will find the following symbols which are designed to help you get the most out of this book:

 An important requirement or point

 An essential requirement, good idea or suggestion

 Note: Used to provide further amplification or information.

Also within the text:

- **Author's Notes**: these are used to provide further amplification or information.

 Author's Note

- *Italic text* indicates a direct quotation (or précised quote) from an ISO 9000 standard, Guidance Note, International or National Standard, etc.
- **Shaded boxes** are used in Chapter 5 and Annex A to show either the full text of ISO 9001:2015 'legal requirements' or a paraphrased version of its requirements.

Disclaimer

Material provided by this service is provided 'as is' without warranty of any kind, either expressed or implied. Every effort has been made to ensure accuracy and conformance to standards available at the time of publication. The user assumes the entire risk as to the accuracy and use of this material. This material may be copied and distributed subject to the following conditions:

- *'Ray Tricker, ISO 9001:2015 in Brief, Published by Routledge'* must be credited as the source of the QMS template;
- this Quality Management System document may not be redistributed for profit. All trademarks acknowledged.

Author's Note

There are three titles in this ISO 9001:2015 series:

ISO 9001:2015 in Brief (this edition)

Now in its 4th Edition, this book is particularly aimed at students, newcomers to Quality Management Systems and the busy executive, with the overall intention of providing them with a user-friendly, very simplified explanation of the history, the requirements and the benefits of the new standard.

Using this book as background material will also enable organisations (large or small) to quickly set up an ISO 9001:2015-compliant Quality Management System for themselves – at minimal expense.

ISO 9001:2015 for Small Businesses (Edition 6)

The new edition of this top-selling Quality Management book includes down-to-earth explanations aimed at helping you to understand what the new standard is all about and to determine what you need to be able to work in compliance with and/or achieve certification to ISO 9001:2015.

It covers all the major changes to this revised ISO standard and how they will affect your work, along with direct, accessible and straightforward guidance that has always been part of this series.

As a bonus, this book also contains a free, customisable example of a complete, generic, Quality Management System that can be adapted to suit any organisation, large or small.

ISO 9001:2015 Audit Procedures (Edition 4)

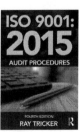

Fully revised, updated and expanded, this 4th edition completes the trio by providing access to methods for auditing against the requirements of ISO 9001:2015.

Although primarily aimed at showing how small business auditors can complete management reviews and internal, external and third party quality audits, this book will prove invaluable to professional auditors.

Containing an overview of the changes made by the 2015 edition of ISO 9001 and how these will affect the way in which audits need to be completed in future, the book also includes an extensive range of audit checklists, explanations and example questionnaires.

Author's Note

Further assistance: for further details about these books and other ISO 9001 consulting services, please email me at ray@herne.org.uk or visit www.thebestqms.com

Chapter 1

What is quality?

 Author's Note

The aim of this short chapter is to introduce the reader to the requirements and benefits of quality, Quality Control and Quality Assurance.

CONTENTS

Why is the word 'quality' (although an everyday word) often misused, misquoted and misunderstood? Probably this is because when most people talk about the quality of an object, or service, they are normally talking about its excellence, perfection or its value. In reality, of course, they should be talking about how much it meets its designed purpose and satisfies the original requirements.

Take for example a £50,000 Mercedes and a £19,000 Ford. It would be very unfair to suggest that the Mercedes is a better-quality car simply because it costs more! Being realistic, both cars meet their predetermined quality requirements because they have been built to exacting standards and are, therefore, equally acceptable as 'quality' vehicles. It is simply that the design purpose and original quality requirements (i.e. the level of quality) differ.

So what exactly is *meant* by the word quality? There are many definitions, but the most commonly accepted meaning of quality is that expressed by The International Organization for Standardization (ISO), namely: *The totality of features and characteristics of products and services that bear on its ability to satisfy stated or implied needs.* In simpler words, one can say that a product or service has **good** quality when it 'complies with the requirements specified by the client'.

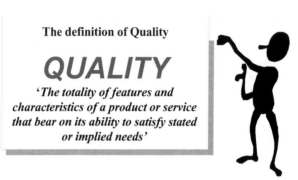

The definition of Quality

QUALITY

'The totality of features and characteristics of a product or service that bear on its ability to satisfy stated or implied needs'

FIG. 1.1 Definition of quality

In other words, quality is based on consumer satisfaction. So in the case of the Mercedes and the Ford, a purchaser of a Mercedes will be satisfied only if they get leather seats, a totally interactive navigation system and surround sound, whereas the Ford driver is happy with crushed velour and a CD player. Their required level of quality differs but each is satisfied with their purchase. The characteristics of each car satisfy customer requirements.

Consumers, however, are not just interested in the level of quality *intended* by the designer, manufacturer or supplier – they are far more interested in the delivery of a product or service which is *consistently* of the same quality. They also want an assurance that what they are buying truly meets the quality standard that was initially offered and/or recommended.

Products and services that are of a consistent quality mean that repeat purchases are more likely – something which any car driver appreciates when considering whether to stay with a preferred make and model.

This consumer requirement has meant that manufacturers and suppliers (especially the larger organisations) have now had to pay far more attention to the quality of their products and services than was previously necessary. Organisations have had to set up proper Quality Management Systems in order to control and monitor all stages of the product or service process, and they have had to provide proof to the potential customer that these will have the guaranteed – and in some cases certified – quality required by the customer. In other words, the manufacturer or supplier has had to work within a Quality Management System to produce their products or deliver their service.

Unfortunately, with the current trend towards micro-miniaturisation and the use of advanced materials and technology, most modern-day products and services have become extremely complex assemblies compared to those that were available just a few years ago. This has meant that many more people are now involved in the manufacture and/or supply of a relatively simple object, and this has increased the likelihood of a design fault occurring.

Similarly, the responsibility for the quality of products and services has also been spread over an increasing number of people, which has meant that the manufacturer's and/or supplier's guarantee of quality has, unfortunately, become less precise.

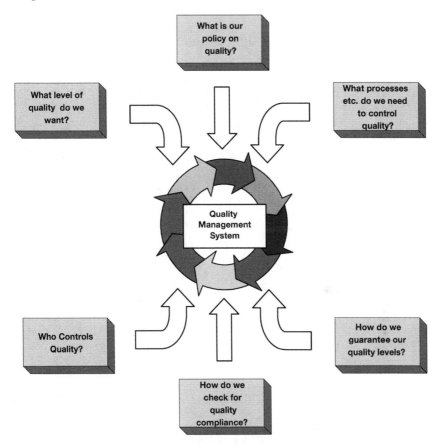

FIG. 1.2 Some of the questions answered by a Quality Management System

The growing demand for an assurance of quality before a contract is awarded has reinforced the already accepted adage that quality products and services play an important role in securing new markets as well as retaining those markets that already exist. Without doubt, in these days of competitive world markets, Quality Assurance has never been more relevant and no longer can suppliers rely on their reputation alone!

Thus the drive towards more quality-led products and services now means that today's major purchasers are not just expecting them to be of sufficient quality, but are also *demanding* proof that an organisation is capable of continually producing quality products or providing quality services. The provision of this proof is normally in the form of an independent third party certification, and this is possibly the single most important requirement for a manufacturer, organisation or supplier.

Until the early 1980s, however, there were no viable third party certification schemes available. But with an increased demand for Quality Assurance during all stages of the manufacturing processes came the requirement for manufacturers to work to recognised standards, and this is why ISO 9000 was first introduced and is now used throughout the world by manufacturers, suppliers and end users.

So in summary, 'quality' *is*:

- a standard which can be accepted by both the supplier *and* the customer;
- giving complete satisfaction to the customer;
- complying consistently with an agreed level of specification;
- providing products and services at an acceptable cost;
- providing products and services that have auditable proof that they are 'fit for the purpose';
- the totality of features or characteristics of a product or service that bear on its ability to satisfy a given need.

Quality *is not* about:

- complying with a specification (as it is possible that the specification may be wrong!);
- being the best (since achieving this ideal may be very costly and could exceed the price that the customer is prepared to pay);
- producing products and services that are 'fit for the purpose' at the point of delivery – as that 'purpose' may be completely different to the customers' actual needs!

FIG. 1.3 Achieving customer satisfaction

1.1 BUT WHO IS RESPONSIBLE FOR MANAGING QUALITY IN AN ORGANISATION?

In the current 2015 edition of ISO 9001, the role of 'Management Representative' (i.e. Quality Manager) has been eliminated as this responsibility has now become:

A management approach centred on quality based on the participation of all its members and aiming at long term success through customer satisfaction and benefits to all members of the organisation and society.

 Which, in other words, means that *everyone* in the organisation is now collectively responsible for the quality of the organisation, its products and its services!

Quality Management ensures that an organisation's products and services are consistent. It has four main components:

- Quality Planning – identifying which quality standards are relevant to the product or service and working out how to satisfy their requirements;
- Quality Control – ensuring that products and services have achieved their highest standard and that their manufacture, installation, modification and/or repair have been completed in an efficient and timely manner;
- Quality Assurance – providing assurance to a customer that the standard of workmanship within a contractor's premises is of the highest level and that all products and services leaving that particular organisation (or supplied by that organisation) are above a certain fixed minimum level of specification;
- Quality Improvement – analysis of performance and systematic efforts to improve it.

Quality Management is *not*, however, just focused on product and service quality, but also on the means to achieve it. Quality Management, therefore, uses Quality Assurance and control of its processes to ensure that their products and services achieve more consistent quality.

1.2 AND SO WHAT IS THE DIFFERENCE BETWEEN A PRODUCT AND A SERVICE?

In the words of ISO (the international standard-setting body promoting worldwide proprietary, industrial and commercial standards):

Product – is an output that is the result of a process that does not include activities that are performed at the interface between the supplier (provider) and the customer.

Service – is the result of a process that includes at least one activity that is carried out at the interface between the supplier (provider) and the customer.

Author's Hint

There are three generic product and service categories: hardware, processed materials and software. Many products and services combine several of these categories. For example, a car consists of hardware (e.g. tyres), processed materials (e.g. lubricants) and software (e.g. engine control algorithms).

1.3 WHAT ABOUT QUALITY CONTROL AND QUALITY ASSURANCE?

We have already defined what quality is all about, but what exactly is meant by Quality Assurance (QA) and Quality Control (QC)?

There seems to be quite a lot of confusion about the meaning of these two topics, and quite frequently people talk about QA when what they actually mean is QC, and although the terms 'Quality Assurance' and 'Quality Control' are both aimed at ensuring the quality of the end product, they are in fact two completely separate processes.

When you think about it, however, the acronym gives it away really as the 'C' in QC is all about *controlling* quality via inspections, checks and tests. The 'A' in QA on the other hand, is about assuring quality and involves working out what is required to guarantee quality will be achieved, and to set out processes, standards, procedures and/or policies to do that.

1.3.1 Quality Control

Definition of Quality Control:

Part of quality management focussed on fulfilling quality requirements.

ISO

'Quality'	Verifies the quality of the output
'Control'	The ability to distinguish what is 'good' (i.e. meets requirements) and what is 'bad' (doesn't meet requirements)
'Quality Control'	The operational techniques and activities that are used to fulfil requirements for quality.

Quality Control (QC) is the amount of supervision to which a product is subjected to, in order to ensure that the workmanship associated with that product meets the quality level required by the design. In other words, it is the control exercised by the organisation to certify that *all* aspects of its activities during the design, production, installation *and* in-service stages are to the desired standards.

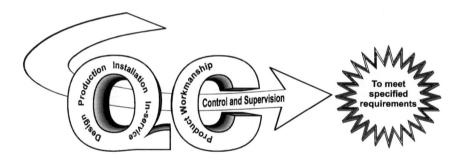

FIG. 1.4 Quality Control

QC is primarily aimed at the *prevention* of errors. Yet, despite all efforts, it remains inevitable that errors will be made. Therefore, the control system should have checks such as the following to detect these:

- what error was made?
- where was it made?
- when was it made?
- who made it?
- why was it made?

Only when all these questions are answered can proper action be taken to correct the error and prevent the same mistake being repeated.

QC is exercised at all levels and as all personnel are responsible for the particular task they are doing, they are *all* Quality Controllers to some degree.

1.3.2 Quality Assurance

The definition of Quality Assurance is:

> *The assembly of all planned and systematic actions necessary to provide adequate confidence that a product, process, or service will satisfy given quality requirements.*
>
> *ISO*

'Quality'	Fitness for intended use
'Assurance'	A declaration given to inspire confidence in an organisation's capability
'Quality Assurance'	A declaration given to inspire confidence that a particular organisation is capable of consistently satisfying needs as well as being a managerial process designed to increase confidence.

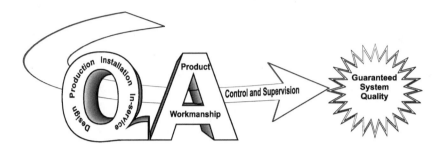

FIG. 1.5 Quality Assurance

QA is also a declaration given to inspire confidence that a product or service has achieved the highest standards and that its production, installation, modification and/or repair have been completed in an efficient and timely manner.

The purpose of QA is:

- to provide assurance to a customer that the standard of workmanship within a contractor's (or subcontractor's) premises is of the highest level and that all products and services leaving that particular organisation are above a certain fixed minimum level of specification;
- to ensure that production and/or service standards are uniform between an organisation's departments or offices and that these remain constant despite changes in personnel.
- In a nutshell, QA is concerned with:
  - an agreed level of quality;
  - a commitment within an organisation to the fundamental principle of consistently supplying the right quality product;
  - a commitment from a customer to the fundamental principle of only accepting the right quality product;
  - a commitment within all levels of contractor and/or customer to the basic principles of QA and QC.

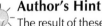 **Author's Hint**

The result of these actions should ideally be checked by someone independent of the work.

In the the case of special projects, customers may require special QA measures or a Quality Plan (See section 2.6).

 Author's Note

Having defined quality, quality control, quality assurance and the difference between products and services, we will now have a look at what a Quality Management System is all about.

Chapter 2

What is a Quality Management System?

CONTENTS

> ### Author's Note
>
> A Quality Management System is exactly what it says on the tin – a system that is capable of managing an organisation's quality requirements enabling them to deliver quality products and services that are fully compliant with the relevant Regulations and their Customers' requirements.
>
> In this Chapter, I shall cover all of the significant elements that make up a Quality Management System and explain what goes into a Quality Policy, Process or Procedure and a Work Instruction. We shall then look at who is responsible for quality in an organisation and how quality is essential to the life cycles of a product or service, what are the purchasers' and suppliers' responsibilities with regard to quality, and finally, how computer technology assists the implementation of Quality Management.

ISO states that a Quality Management System is:

A set of interrelated or interacting elements to establish the overall intentions and direction of an organisation related to quality as formally expressed by Top Management and to achieve those objectives.

ISO 9000

Or to put it another way, it is the structure, policies, procedures, processes and resources needed to implement a quality managed structure of organisational responsibilities, activities, resources and events that together provide processes focused on achieving Quality Policy and quality objectives to meet customer requirements. Thus a company controls its business through the application of a *business* management system, *not* – as the previous ISO 9001:2008 standard suggested – through a Quality Management System. So one mustn't get too hung up on the word 'quality'!

The current ISO 9001:2015 sets the standard for a business management system which, if implemented correctly, will lead to products and services of a predetermined quality, which will in turn satisfy the customer's requirements and expectations.

Author's Hint

In previous editions of this title, I suggested that a better name for ISO 9001 would possibly be '*A Standard for Business Management Systems*', especially as there are so many other management systems directly related to ISO 9001. Indeed, a number of companies that have implemented ISO 9001 have deliberately avoided using the term 'quality', preferring instead to simply call their business activities a 'management system'.

Following recent international agreement, however, that in future *all* management system standards shall have the same high-level structure, identical core text, as well as common terms and definitions, this means that business management is paramount and perhaps Quality Management will become more of a subset to the former.

Having said that, for the purposes of this book, we will stick with the term Quality Management System, as it is ill advised to upset the International Standards Organization at this stage!

So what is meant by a successful Quality Management System (QMS)? To be effective, an organisation relies on a variety of interactions and inputs (as indicated below) in order to achieve their ultimate goals for Quality Assurance (QA) and Quality Control (QC).

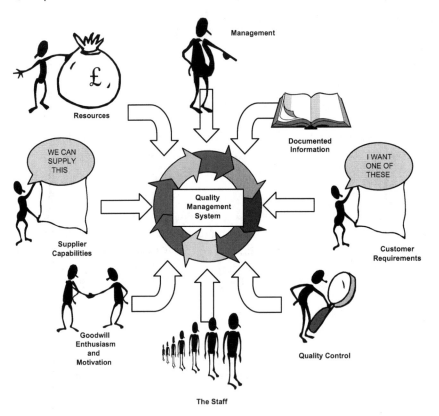

FIG. 2.1 The ingredients of a successful Quality Management System

Together, these documents comprise the organisation's Quality Management System and describe their capability for supplying products and services that will comply with laid-down quality and other regulatory standards.

Previously this was contained in the 'Quality Manual', which was a document setting out the quality policies, systems and practices of an organisation's Quality Management System.

With the introduction of ISO 9001:2015, the International Standards Organization (ISO) Committee responsible for updating the standard also created a standard format (i.e. Annex SL – see Chapter 5.5) with the same high-level structure, identical core text and common terms and definitions – *but* they *left out* the requirement for an organisation to possess a formal Quality Manual – preferring to call everything associated with an organisation's documents and records 'documented information'.

Whereas ISO 9001:2008 used specific terminology such as 'document' or 'documented procedures', 'Quality Manual' or 'Quality Plan', the 2015 edition of this International Standard now contains requirements to 'maintain documented information'.

However, ISO – in acknowledging that there are over a million certified ISO 9001:2008 companies currently using a Quality Manual, etc. – have stated (in Annex A to ISO 9001:2015) that

> There is <u>NO</u> requirement in this International Standard for its structure and terminology to be applied to the documented information of an organisation's Quality Management System!

So an organisation can choose whatever they want to call the document included in their quality processes and procedures, etc. You can call it a 'Quality Manual' or just retain it as 'documented information' – you won't be breaking any rules!

Author's Hint

Thus to avoid any confusion, I have continued to use the term 'Quality Manual' in this book and leave it up to my readers to choose what to call their 'database', be it Quality Manual, Management Manual or simply an Organisational Management Manual.

All that aside, in order to be successful an organisation must be able to prove that it is capable of producing products and services that meet the customer's complete satisfaction and which not only conform to the customer's specific requirements but are always of the desired quality. An organisation's QMS is, therefore, the organisational structure of responsibilities, procedures, processes and resources for carrying out Quality Management.

As such, it must be planned and developed so that it is capable of maintaining a consistent level of QC.

The 'quality loops' shown in Figs 2.2 and 2.3 should always be followed by an organisation to ensure that all aspects of the production and supply cycles have been considered in the QMS.

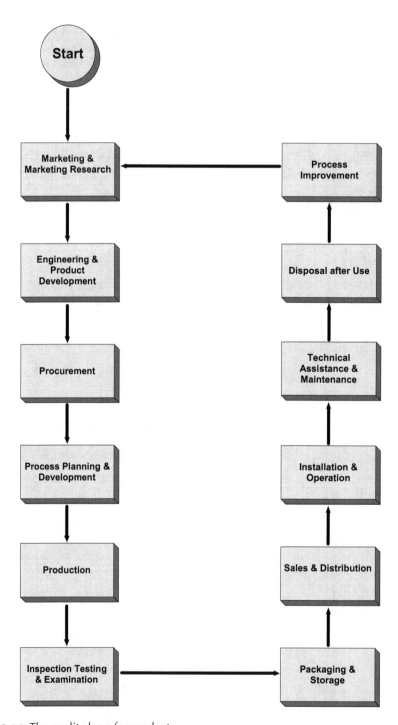

FIG. 2.2 The quality loop for products

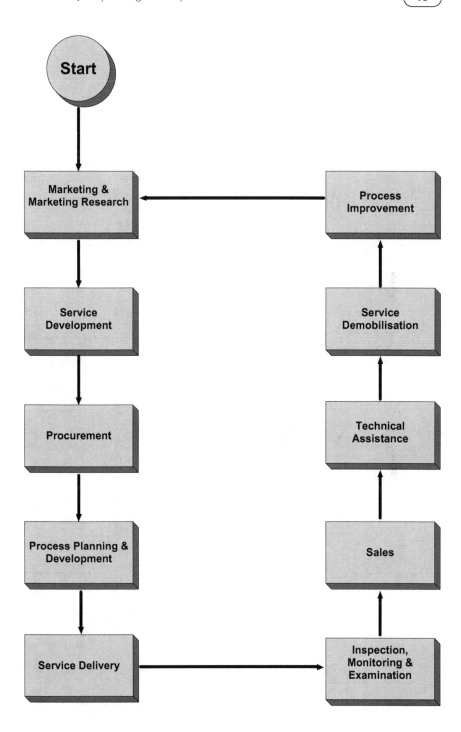

FIG. 2.3 The quality loop for services

So whether you produce 'nuts and bolts', design software or provide a service (such as public relations), a QMS is ideal for running your organisation.

However, to be effective, the QMS must be structured to the organisation's own particular type of business and should be flexible enough to consider all functions such as customer liaison, design, purchasing, subcontracting, manufacturing, training, installation, updating of QC techniques and the accumulation of quality records. As touched on before, in most organisations this sort of information will normally be found in its Quality Manual.

The type of QMS chosen will, of course, vary between one organisation and another, depending upon its size and capability. There are no set rules as to exactly how these documents should be written. Nevertheless, they should – as a minimum requirement – be capable of showing the potential customer exactly how the organisation is equipped to achieve and maintain the highest level of quality throughout the various stages of design, production, installation and servicing.

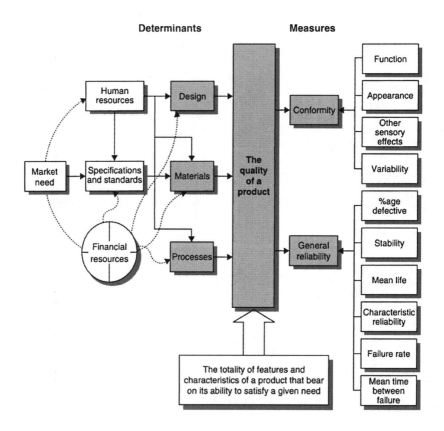

FIG. 2.4 Some of the determinants and measures of the quality of a product

As an example, some of the determinants and measures of the quality of products and services are shown in Figs 2.4 and 2.5.

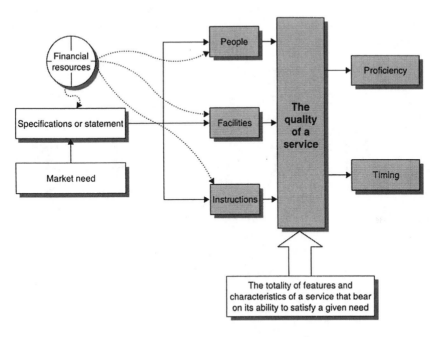

FIG. 2.5 Some of the determinants and measures of the quality of a service

Note: Figs 2.4 and 2.5 are extracts from BS 4778:1979, which have been reproduced with the kind permission of BSI. Although the 1979 edition has been superseded, these figures are included here since they still illustrate the concept.

2.1 WHAT ARE THE REQUIREMENTS OF A QUALITY MANAGEMENT SYSTEM?

To be successful, an organisation (whether large or small) *must*:

- be able to offer products and services that satisfy a customer's expectations;
- agree with the relevant standards and specifications of a contract;
- be available at competitive prices;
- be able to supply products and services at a cost that will still bring a profit to that organisation.

Organisations must, above all, provide quality products and services that will promote further procurement and recommendations.

So, how can your organisation become a Quality organisation?

Well, it is *not* just a case of simply claiming that you are a reliable organisation and then telling everyone that you will be able to supply reliable products and services! Nowadays, especially in the European and American markets, purchasers are demanding proof of these claims: proof that *you* are the organisation that *they* should be dealing with.

FIG. 2.6 No proof of quality = no business!

How can anyone supply this proof? Well, the easiest and most recognised/ generally accepted way is to work in conformance with the requirements of ISO 9001:2015. This standard provides guidelines for organisations wishing to establish their own QMS and thereby control the quality of their organisation – from within their organisation.

FIG. 2.7 The benefits of proof

But it doesn't just stop there! Sometimes a contract will require an organisation to comply with the specifications of other standards. For example, a British semiconductor component manufacturer would be required to meet the mechanical standards contained in BS EN 60191–6–22, or for a dental laboratory it could be the European Community Council's 'Medical Device Directives' (as detailed in EN ISO 13485), and so a well-structured QMS can prove extremely useful when dealing with these situations.

As previously mentioned, an organisation must *prove* their 'organisation's capability' by showing that it can operate to a fully auditable and compliant QMS,

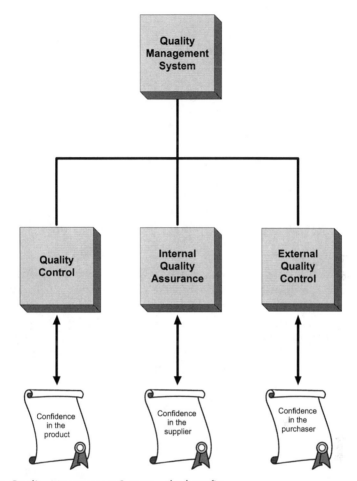

FIG. 2.8 Quality Management System – the benefits

and Fig. 2.8 shows how a QMS benefits an organisation by providing both that organisation and their potential customers with the necessary proof.

To satisfy these requirements, an organisation's QMS has to encompass all the different levels of Quality Control and Quality Assurance that are required during the various stages of design, production and acceptance of a product or a service (i.e. system or process), and be capable of guaranteeing quality.

These requirements generally cover the following topics:

- organisational structure;
- measurement of QA;
- contract specification;
- design control;
- purchasing and procurement;

- production control;
- product and service testing;
- handling, storage, packaging and delivery;
- after-sales service;
- risk analysis.

2.2 WHAT ARE THE COSTS AND BENEFITS OF HAVING A QUALITY MANAGEMENT SYSTEM?

The main requirement of a Quality Management System is to:

> *possess a system by which an organisation can aim to reduce and eventually eliminate nonconformance to specifications, standards and customer expectations in the most cost effective and efficient manner*
>
> *ISO*

In practice, some QA programmes can be very expensive to install and operate, particularly if inadequate QC methods were used previously. If the purchaser requires consistent quality then they must pay for it, *regardless* of the specification or order which the organisation has accepted. Yet, against this expenditure must always be offset the savings in scrapped material, rework and general problems arising from lack of quality. How much an organisation benefits from its QMS is directly related to the money it invests in quality.

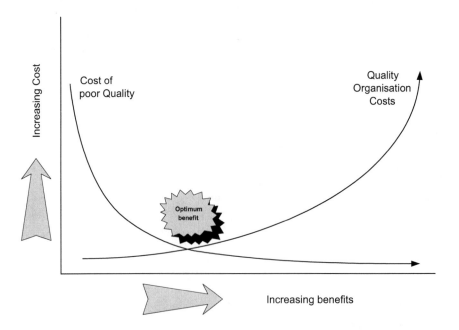

FIG. 2.9 Quality Management System costs

However, it is always possible to put *too* much money into QC, and so the optimum benefit comes when the investment in QC is balanced against the most significant reduction in the cost of poor quality. As can be seen from Fig. 2.9, any further investment beyond this point will not result in substantial gains.

The main benefits of Quality Management are:

- an increased capability to provide products and services that consistently conform to agreed specifications;

FIG. 2.10 An increased capability

- a reduction in administration and production costs because of less wastage and fewer rejects;

FIG. 2.11 Reduced costs

- a greater involvement and motivation within an organisation's workforce;

FIG. 2.12 Motivated staff are happy

- improved customer relationships through fewer complaints, thus increasing sales potential.

FIG. 2.13 Customer satisfaction

For an organisation to derive any real benefit from a QMS, everyone in the organisation must:

- fully appreciate that QA is absolutely essential to its future;

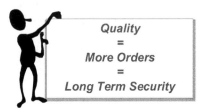

FIG. 2.14 Understanding the reason for quality

• fully appreciate that QA is absolutely essential to their future;
Quality = More orders = Long term security

- know how they can assist in achieving quality;

• know how they can assist in achieving quality;

Why not ... ?

FIG. 2.15 How your staff can assist

• be stimulated and encouraged to do so.

- be stimulated and encouraged to do so.

You're a winner with Quality!

FIG. 2.16 Be encouraged by the organisation

With an effective QMS in place, the organisation will achieve increased profitability and market share, and the purchaser can expect reduced costs, improved products and services, fitness for role, increased satisfaction and, above all, growth in confidence.

But, *without* an effective QMS, organisations will definitely suffer.

Author's Hint

In the following sections we shall look at each element that makes up a QMS and in what way they can combine to clearly define how a business achieves its goals.

2.3 WHAT IS A QUALITY MANUAL?

Since the first edition of ISO 9001, and even before that, when it was referred to by a different number (see chapter 3), it has been an accepted fact that a Quality Manual is '*a document specifying the Quality Management System of an organisation*'.

A Quality Manual is recognised as an official document produced by an organisation that details how its Quality Management System operates. A typical Quality Manual will include the company's Quality Policy and goals, as well as a detailed description of its QC system including Staff roles and relationships, procedures, systems and any other resources that relate to producing high-quality goods or services.

Author's Hint

As mentioned earlier in this chapter, with the introduction of ISO 9001:2015 the ISO Committee responsible for updating the standard removed the requirement for an organisation to possess a formal Quality Manual.

As the quality processes and procedures, etc. of an organisation are all interrelated, there must be some sort of centralised library for these documents (be they hardcopy or digitised copies) and surely the most logical place for these is in a file called the 'Quality Manual' – a document that has existed and has been used by auditors and Quality Managers for more than 50 years!

At the time of writing this book, the 'jury' still appears to be out on this 'ruling' and many organisations (particularly those already ISO 9001:2008 certified) intend to continue using their existing, albeit slightly modified, Quality Manual so as to include the changes made by ISO 9001:2015.

Whatever the final ruling regarding the title of this particular document, it will become the formal record of its QMS and it will be:

- a rule book by which an organisation functions;
- a source of information from which customers may derive confidence;
- a means of defining the responsibilities and interrelated activities of every member of the organisation;
- a medium for defining the level of quality that an organisation wishes to consistently deliver;
- a vehicle for auditing, reviewing and evaluating the organisation's QMS.

There are no set specific rules concerning what format and structure you should use to produce your Quality Manual, provided that it:

- includes a statement of your organisation's policy towards Quality Control;
- contains details of your organisation's Quality Management structure together with job descriptions and responsibilities;
- describes your organisation's QC requirements, training programmes, etc.

You can use one or many formats, from simple checklists and flowcharts to descriptive text. You can use any type of media and so a Quality Manual can be in hardcopy, softcopy, online (via your business website or cloud-based), as an intranet help file or via one of many other types of IT systems we currently have available today.

There is no limit to the size of the Quality Manual. It can be a simple pamphlet with a reference index to other stored documents (as indicated in the previous paragraph) or it can be a complete library.

Whatever the size and format of your Quality Manual, it still remains as the single point of reference required to run all aspects of your organisation to consistent quality levels.

It is the heart of a QMS and is essential for anyone considering applying for ISO 9001:2015 certification.

2.3.1 What goes into a Quality Manual?

FIG. 2.17 What goes into a Quality Manual?

The answer to 'what goes into a Quality Manual' is virtually anything and everything, and depends on what you need in order to run your business efficiently. For example, a 'Mercedes' sort of organisation with many different departments might have more than one departmental Quality Manual supporting a company-wide Quality Manual, but a small to medium-sized business will probably only need a small manual for their QMS.

Whatever the size of your organisation and whether you produce, install or merely use a product or service, the overall purpose of the Quality Manual is to act as your organisation's overarching policy 'bible', which (if required) will identify subsets of Quality Processes, Quality Plans, Quality Procedures and Work Instructions and provide templates of the various forms and documents used by

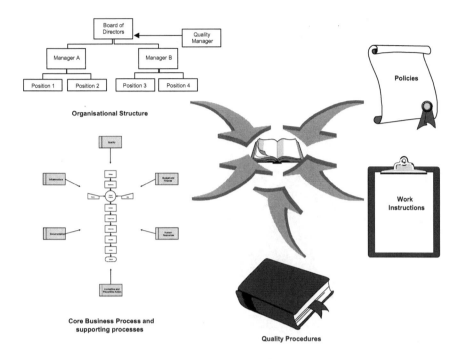

FIG. 2.18 The basic contents of a Quality Manual

your organisation – such as production control forms, inspection sheets and documents used to purchase components from your subcontractors.

Quality Procedures and Work Instructions will include details of the specifications that must be complied with. For a producer these may include:

- particulars of drawings;
- supporting documentation;
- tools and gauges that are going to be used;
- sampling methods;
- any tests which have to be made;
- test specifications and procedures;
- the acceptance/rejection criteria, etc.

For organisations providing a service, the following may be found in their Quality Manual:

- response time criteria;
- service standards;
- customer satisfaction and complaints procedures;
- courtesy requirements (e.g. acceptable telephone manner).

Note: For a complete description and guidance on how to develop a Quality Manual, the reader is referred to ISO 10013 – 'Guidelines for Quality Management System documentation'.

2.3.2 What does each part of the Quality Manual do?

Each part of a Quality Manual has a specific role to play, as shown below.

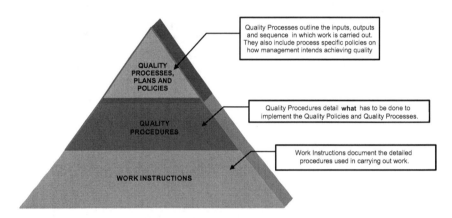

FIG. 2.19 What each part of a Quality Manual does

Author's Hint

Most organisations use their Quality Manual to address each clause of ISO 9001:2015. Although this is *not* an ISO requirement, it is not a bad idea as it will assist internal and external auditors when they come to evaluate an organisation's QMS for compliance with the standard.

2.4 WHAT IS A QUALITY POLICY?

A Quality Policy is a statement of an organisation's overall quality intentions and the direction of an organisation regarding quality which is formally expressed by Top Management. It outlines *how* management intends to achieve quality and dictates how every other aspect of an organisation's QMS is set up and run.

There are three types of policy statement that an organisation needs:

- **Mission Statement**: A very brief, high-level statement of intent from senior management;
- **Corporate Policy Statement**: An expansion upon the Mission Statement;
- **Process-specific policies**: Clear statement of intent for each process performed by a business.

2.4.1 Mission statements

Whilst ISO 9001:2015 does not directly call for a mission statement, it is ideal as a means of starting a QMS as it is aimed at polarising the mind and pointing everyone within a business in one particular direction.

This high-level policy statement should be focused on customer satisfaction and:

- be appropriate for the needs of the organisation and its customers;
- involve everybody within the organisation;
- outline the organisation's goals and objectives;
- be communicated and implemented throughout the organisation;
- be understood by everyone.

Mission statements should ideally be short and capable of being instantly memorised, e.g.:

We, the LUR company, aim to deliver effective, evidence-based management products and systems in accordance with the recommendations of ISO 9001:2015 and other industry equivalent standards.

FIG. 2.20 A commitment from senior management

All these mission statements have certain things in common, in that they:

- define a goal for the business;
- state corporate values;
- start to define measurable objectives;
- imply a need for employee competence;

- address risk management;
- address customer satisfaction;
- provide a lead towards continual improvement.

These points should be considered when writing your own mission statement.

2.4.2 Corporate Policy Statement

In a similar way, if an organisation is serious about setting up a QMS, then Top Management must commit itself by stating their policies on quality. Without a firm commitment from senior management, an organisation's QMS will fail. This Corporate Policy Statement (CPS) is usually achieved by management putting together a document that outlines their intentions, based upon their mission statement.

This high-level policy statement should be focused on customer satisfaction and:

- be appropriate for the needs of the organisation and its customers;
- involve everybody within the organisation;
- outline the organisation's goals and objectives;
- be communicated and implemented throughout the organisation;
- be understood by everyone.

ISO 9001:2015 unintentionally gives us some guidance in putting together a policy statement, in so much as the standard is based upon seven Quality Management Principles (see Chapter 5.7). By establishing your policy against these principles you should end up with a robust statement. Box 2.1 shows how such a policy statement may look.

☀ Author's Hint

ISO 9001:2015 makes no reference to any financial policies, which I feel is a bit of an oversight bearing in mind that all companies strive to maximise profits and minimise overheads. So you may want to consider adding an eighth statement to the effect that you will visibly demonstrate cost effectiveness.

2.4.3 Process-specific policies

To comply with the requirements of ISO 9001:2015, you will need to determine, establish, implement, maintain and continually improve written processes for your QMS and, in particular, the design and development of your products and services.

The actual number of process-specific policies that you will need depends entirely

FIG. 2.21 A clerk's nightmare

upon what activities your particular organisation needs to control. It may be that all you want to do is instill some discipline into your mailroom because the clerks are sinking under a mountain of paper!

Box 2.1

The policy of this company is to achieve and maintain a high standard of management in all aspects of our operation and to continually satisfy the expectations of our customers.

We will conduct our business through the following principles:

- **Customer focus**: We will deliver products and services that comply with our predetermined requirements which recognise the needs and expectations of our customers.

- **Leadership:** We will provide direction for the business by establishing clear objectives that serve to fulfil business goals.

- **Engagement of people**: We will develop our business through the involvement of our Staff, by utilising their knowledge and experience.

- **Process approach**: We will run our business through a structured process-based management system and will ensure that all our Staff work to the requirements of this system.

- **Improvement**: We will commit to enhancing our management system through the proactive identification and implementation of improvement opportunities.

- **Evidence-based decision making**: We will make informed business decisions using the analysis of data obtained from suitable metrics.

- **Relationship management**: We will develop relationships with our suppliers and work with them to improve sub-contractor performance.

Signed _____
Managing Director

At the other end of the scale, your organisation may be looking to control all aspects of its work. It is your decision, but whatever the reason you must have policies for each process that can be clearly dictated in regard to the main activities of your organisation, thereby avoiding any ambiguity.

FIG. 2.22 No policies – no Quality Control!

To start getting things under control, you will need to break down these activities into a series of inter-related processes that describe how they manage your quality. Then once the Core Business Process and its Supporting Processes have been identified, it is then relatively straightforward to define policies for each of the processes. These processes will then probably need to be supported by a number of QPs. You are, of course, at liberty to set down in words as many procedures as you feel appropriate, remembering that procedures are generally used to amplify the detail shown on process maps.

Author's Hint

Just a word of warning; no one likes to read procedures on how to do things. Process mapping is the better option as it visually represents your business and so it is best to keep written procedures to a minimum and only use them when they are either mandatory (to your particular business) or will add value and clarity to a process map.

'A picture is worth a thousand words'!

FIG. 2.23 A picture paints a thousand words!

2.4.4 How to set quality objectives

As ISO 9001:2015 is quick to tell us, quality objectives are established to provide a focus to direct your business with the overall intention to lay down the desired results you wish to achieve, thereby enabling you to apply resources to achieve these goals.

Quality objectives need to be consistent with the high-level corporate Quality Policy and in turn relate to the process-specific policies.

There is one basic rule that needs to be applied when setting quality objectives, that being, think 'S.M.A.R.T.'

Specific
Measurable
Attainable
Realistic
Timely

FIG. 2.24 Being SMART

Objectives must be concise and to the point. Wishy-washy statements such as 'we will be the best' are unrealistic as there is no way of telling whether you are the best in the business.

A statement such as 'We aim, by the end of this fiscal year, to deliver 99.8% of all our products and services within 48 hours of receipt of order' meets the S.M.A.R.T. criteria as follows:

- **Specific**: The objective relates to a physical activity, in this case the delivery of a product or a service. A specific objective is one that can be observed and therefore verified against predetermined criteria. In the above example 'delivery' is the observable action and 'within 48 hours' is the specific criterion.
- **Measurable**: The objective is capable of being measured, as targets have been set through percentages and timescales. A mechanism can therefore be put in place to capture these data, be it a dispatch note that records departure and arrival times of the delivery van or, in this high-tech world, a satellite tracking system which automatically records when the van arrives at its destination. In essence you need a dependable system that can measure your success in achieving the objective.
- **Attainable**: Clearly this company is confident that it can deliver 98% of its products and services on time; therefore the objective is, in its opinion, attainable. Ideally such an objective should be agreed by the staff that are actually performing the activity, as if you have only provided them with a delivery bike with a flat tyre, they may not agree that 98% is attainable! The target must be sufficiently demanding to ensure the delivery team are not wasting time and therefore becoming inefficient.
- **Realistic**: This must not be confused with 'Attainable'. Realism in this context means whether the objective is realistically applicable to the people responsible

for performing the task. In the example above you would not ask the delivery team to increase profits by 5% over the next fiscal year. However, by asking them to deliver 98% of their products and services within 48 hours of receipt of the order, this should lead to customer satisfaction and therefore repeat orders, which in turn should improve the financial bottom line. In short, apply specific objectives to those who are capable of delivering them.

• **Timely**: No great surprise here. There is no point in setting an objective without specifying a time by which it should be achieved. Our delivery company has given themselves until the end of the fiscal year to attain their objective.

Objectives that are S.M.A.R.T. can be monitored. In this way you will be able to establish whether you have been successful in achieving your goals, or conversely, if you have failed to meet them you will be able to quantify the failure rate, identify ways in which to improve and ultimately set new objectives.

2.4.5 How to go about setting measurable objectives

Simply ask yourself what influences the way in which a process is performed. These 'influences' are the known variables that will either make the process succeed or fail.

For example, brewing beer is influenced by the type of hops, the yeast, temperature, brewing time, shelf life, etc. All these influences can be considered as success criteria. So if your objective is to make consistently acceptable barrels of beer, then it is not difficult to see that it is important to measure the criteria that impact upon your success as a brewer.

FIG. 2.25 Measuring success criteria leads to meeting objectives

2.4.6 Fancy a RUMBA?

For those of you with a love for acronyms, you may like to consider the alternative to S.M.A.R.T., that being 'R.U.M.B.A'. Whichever way you choose to generate your objectives, the same principles apply.

Reasonable
> You are capable of meeting the requirement within current rules, regulations and legislations, i.e. it will not put you behind bars!

Understandable
> You understand what the objective is aimed at achieving.

Measurable
> The objective can in some way be monitored to see that it is meeting its intended target.

FIG. 2.26 Fancy a RUMBA?

Believable
> Your Staff feel that it is a realistic objective and will therefore strive to meet it.

Achievable
> The objective is theoretically possible.

2.5 WHAT IS A *QUALITY* PROCESS?

ISO defines a Quality Process as 'a system which uses resources to transform inputs into outputs'.

FIG. 2.27 The basis of a Quality Process

Processes can be found all around us. Take for example the process of getting up in the morning. The initial starting point (input) would be you sleeping in bed. The alarm clock would carry out the process of waking you up, which would be followed by the action of getting up (the output). The alarm clock (the process)

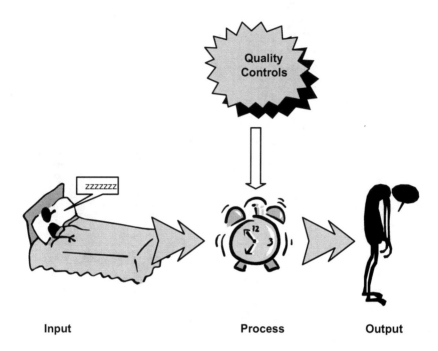

Input   **Process**   **Output**

FIG. 2.28 The simple process of waking up!

would then need to have some Quality Controls (such as checking that the clock is telling the correct time before going to bed, setting the wake-up time and ensuring the alarm is switched on) to ensure that the process will work.

An example of a more complex process would be building a computer network, where inputs are both human and physical. The method of combining the skills of the technician, the component parts of the system and, most importantly, meeting the customer's requirements, are controlled by a process specifically developed to deliver an acceptable result (i.e. the output).

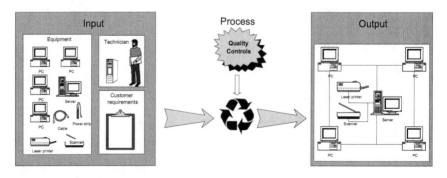

FIG. 2.29 A more complex process

2.5.1 The process approach

Any activity that receives inputs and converts them to outputs can be considered as a process. Often, the output from one process will directly become the input into the next process.

For organisations to function effectively, they will have to identify and manage numerous interlinked processes. This systematic identification and management of the processes employed within an organisation (and particularly the interactions between such processes) is referred to as the 'process approach'.

In addition to inputs, outputs and Quality Controls, processes include physical attributes like suppliers, customers and the resources needed to perform the process. A more accurate depiction of a process is shown below.

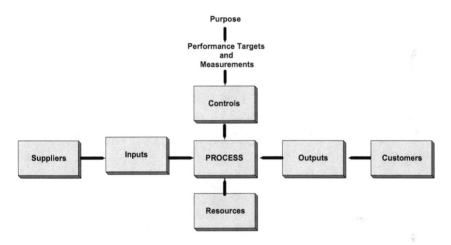

FIG. 2.30 The composition of a process

A brief explanation of each process component is given below:

- **Purpose**: A short description of what the process is intended to achieve, i.e. what value it is adding to the business.
- **Performance targets and measures**: the predetermined performance levels you decide and the tangible metrics that are used to prove the process is performing.
- **Process owner**: Every process should have one designated owner who is accountable for its execution and its continued applicability.
- **Suppliers**: Consider this the source of inputs to a process. It could be another department or another process.
- **Input**: Those items needed to produce the outputs. They should not be confused with the resources needed to perform the process. For example, a set of accounts could be the input needed for performing a financial analysis process.

- **Controls**: Those standards defined by yourself (or possibly legislation) that impact upon the way a process is performed.
- **Resources**: These can be:
  - **Personnel**: Define who is responsible for carrying out the process.
  - **Infrastructure**: Any equipment required to perform the process to an acceptable level.
  - **Support systems**: These could be computer software packages required to perform a specific activity.
- **Outputs**: The expected deliverable from a process. This 'may' form the input to another process, but in all instances it will add value to a business.
- **Customers**: Ultimately all processes deliver something (the output) to a customer, who could be the ultimate purchaser of a particular product or service or someone internal to your own business, such as a stores controller taking delivery of equipment from goods inward.

All the above points should be considered before embarking upon drawing up a process.

In ISO 9001:2015, Core Business, Primary Supporting and Secondary Supporting processes are used in an identical way to define how resources and activities are combined, controlled and converted into deliverables. Processes are the key to providing a clear understanding of what an organisation does and the QCs it has in place to do those activities.

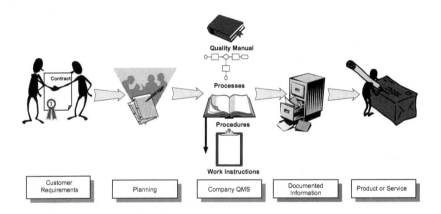

FIG. 2.31 The basic Quality Processes

These processes are explained in more detail in the following pages.

2.5.2 Core Business Process

The Core Business Process describes the end-to-end activities involved in producing a contract deliverable or marketing opportunity. It commences with

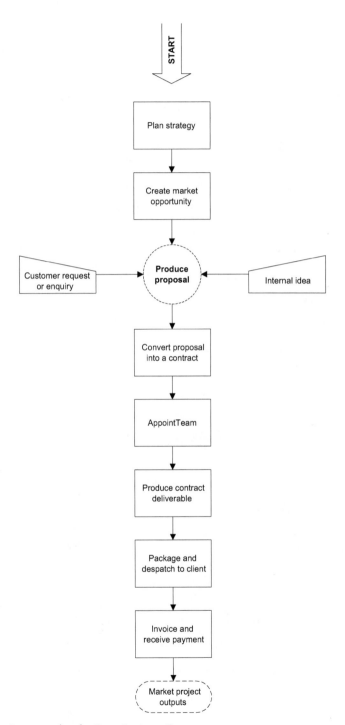

FIG. 2.32 An example of a Core Business Process

the definition of corporate policy and ends when the product or service is manufactured and/or marketed, or the service completed.

 Note: All aspects concerning the products and services (together with their associated marketing activities) should be retained as documented information and used to plan future strategies.

Within reason, anything can be added to make the process more efficient, but nothing can be eliminated and when the Core Business Process states that certain tasks must be performed in sequence, then it must be reflected in the implementation. In the same manner, any specified formulas or steps associated with a task must also be reflected within the implementation.

2.5.3 What types of Core Processes would a business require?

If you are a small or medium-sized business, then probably five core processes will suffice, namely:

- Sales & Marketing
- Accounting & Technology
- Quality & Products or Service Delivery
- Management, HR & Finance
- Product or Service Development.

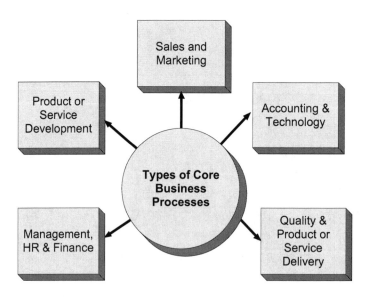

FIG. 2.33 Types of Core Business Process

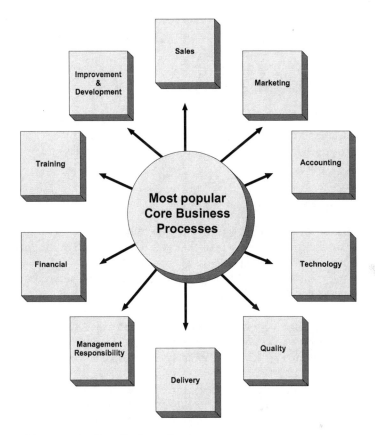

FIG. 2.34 Most popular Core Business Processes

When you start to grow, however, this will introduce new complexities that might require more employees and more focus, so these five processes will have to be expanded (e.g. sales and marketing might have to become two separate entities) and the number of core processes required could increase to ten or even more. There is no restriction to how many primary business processes an organisation can have, but you will find that the ten listed below are the most commonly used today.

- Customer Acquisition (Sales)
- Customer Strategy & Relationships (Marketing)
- Accounting Management
- Technology Management
- Quality, Process Improvement & Change Management
- Products/Service Delivery
- Management Responsibility
- Financial Analysis, Reporting, & Capital Management

- Training
- Product and Service Improvement & Development.

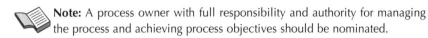

Note: A process owner with full responsibility and authority for managing the process and achieving process objectives should be nominated.

The Core Business Process may then be supplemented (depending on the size of the business) by a number of supporting processes that describe the infrastructure required to manufacture (or supply) the products and services on time.

2.5.4 Supporting processes

Of course the only way for an organisation to ensure repeat orders is to control quality. Consequently, it is essential that you define your Quality Policy and objectives for each supporting process.

Thus, for each process within the flowchart there will be accompanying documentation detailing:

- **Objective:** what the process aims to achieve;
- **Scope:** what the process covers;
- **Responsible owner:** who is ultimately responsible for implementing the process;
- **Policy**: what the organisation intends on doing to ensure quality is controlled;
- **Key performance indicators:** those items of objective evidence that can be used as a way of monitoring performance of the process;
- reference to supporting system documentation (i.e. QPs and WIs).

All businesses revolve around taking inputs and putting them through a series of activities that turn them into useful outputs, be they products or services. These activities are the supporting processes.

A flowchart of a typical supporting process is shown in Fig. 2.35.

Secondary supporting processes

In addition to primary supporting processes, there will be *secondary supporting processes* that run in parallel with and support the primary supporting processes. These secondary supporting processes are equally important as they control all other activities that may influence the quality of products and services – as well as supporting and achieving the primary supporting processes.

An example of a secondary supporting process is shown in Figs 2.36 and 2.37. These secondary supporting processes will have an identical structure to the primary supporting processes, and will also have their own associated supporting documentation (i.e. QPs and WIs).

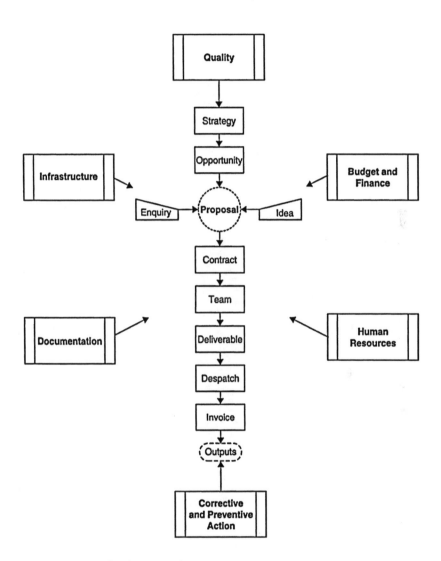

FIG. 2.35 An example of a supporting process

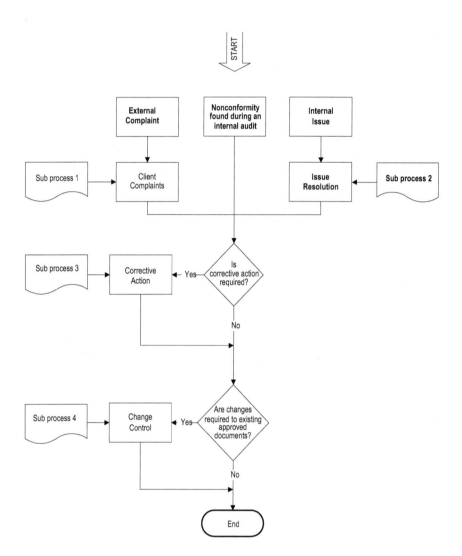

FIG. 2.36 Flowchart showing typical primary supporting process (in this case, Corrective Action)

Identification, provision, management and support of staff

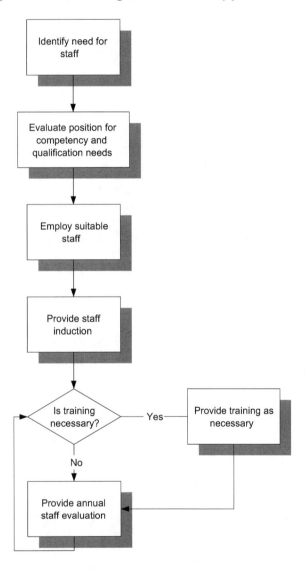

FIG. 2.37 An example of a secondary supporting process flowchart

Secondary supporting processes may include such items as:

- identification and provision of suitable Staff;
- management and support of Staff;
- identification and provision of information;
- identification and provision of materials;
- identification and provision of equipment and facilities;
- management of the QMS;
- continual improvement.

2.5.5 Inter-relationship of process documentation

All processes are documented to give a complete picture of how to perform the activity to a consistent level of quality. The level of detail varies depending whether it is a:

- **Process**: an outline of its objective, scope and key performance indicators;
- **Quality Procedure**: an enlargement of the process explaining how it is controlled;
- **Work Instruction**: the 'fine print' required to perform a specific activity.

All of these documents are explained in more detail elsewhere in this book.

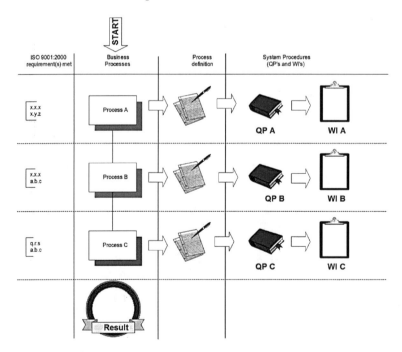

FIG. 2.38 The inter-relationship of documented processes with QPs/WIs

Author's Hint
By using a matrix such as this, it is possible to identify the parts of ISO 9001:2015 which are met by each process.

Throughout ISO 9001:2015, the requirement for continuous improvement is frequently (and heavily) emphasised. The following process model clearly shows how the six major sections (i.e. leadership, planning, support, operations, performance and improvement) of ISO 9001:2015 inter-relate and how the improvement processes continuously revolve around all other aspects of Quality Management.

2.5.6 The hierarchy of processes

When developing process maps, they can become very complex. It is therefore recommended that you develop a hierarchy of process maps, which enables users to 'drill down' through a number of levels to reach the desired level of detail, and Fig. 2.39 shows one way in which this can be achieved.

Level 1 (i.e. The Overview) does nothing but show a summary of the Core Business Process and the supporting processes.

Should you wish to know more detail on a certain process, then by drilling down to the Level 2 (High Level Process Linkages) maps you can get more information on the activities making up that specific process.

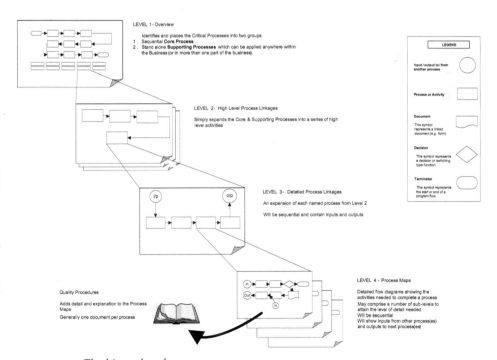

FIG. 2.39 The hierarchy of processes

If more detail is required, then further levels (Level 3 – Detailed Process Linkages) can be added as shown, until you perhaps choose to supplement the maps (Level 4 – Process Maps) by a written procedure.

This type of hierarchy is ideal for producing computer-based management systems, where the functionality offered by websites enables process maps to be linked, thereby simplifying navigation between maps. Further details on using computer technology to enhance your managements system can be found later in this book (see Chapter 2.13).

2.6 WHAT IS A QUALITY PLAN?

A quality plan is a document, or several documents, that together specify quality standards, practices, resources, specifications, and the sequence of activities relevant to a particular product, service, project or contract. They are used to record the quality requirements for their activities and to monitor and assess adherence to their requirements.

You may already have QCs for your normal product or services, but what do you do if someone wants something completely different? You could simply apply your existing QCs, but it is unlikely that they will cover any eventuality. What you need is a Quality Plan to address the specific requirements of that particular contract and which are a true record of the customer's requirements.

A Quality Plan is effectively a subset of the actual Quality Manual. Some may even say that it is a 'customised Quality Manual', as the layout of the Quality Plan will be very similar to that of the Quality Manual and need only *refer* to the QPs and WIs contained in that Quality Manual, supplemented by detailed *contract-specific* QPs and WIs.

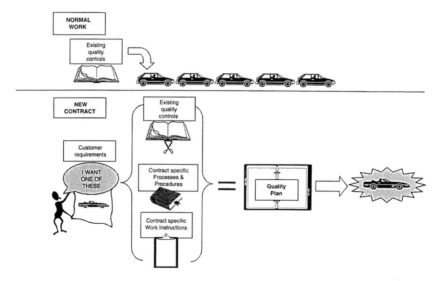

FIG. 2.40 Quality Plans are needed to control the quality of specific projects

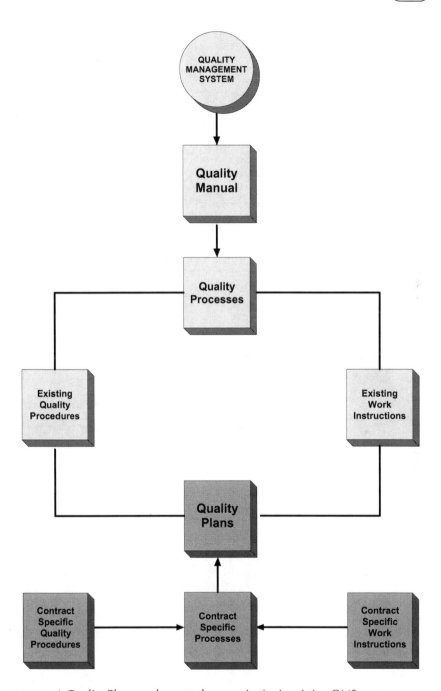

FIG. 2.41 A Quality Plan supplements the organisation's existing QMS

In essence, Quality Plans provide a collated summary of the requirements for a specific project. They will cover all of the quality practices and resources that are going to be used, the sequence of events relevant to that project, the specific allocation of responsibilities, methods, the QPs and WIs, together with details of the testing, inspection, installation (if required) examination and audit programme stages.

In addition to being ideal for controlling the quality of a product, Quality Plans are equally suited to the delivery of processes and/or services. For example, your organisation may provide catering services, in which case Quality Plans would be an ideal way of controlling wedding receptions, etc. because no two events are the same.

The main requirement of a Quality Plan, however, is to provide the customer (and the workforce) with clear, concise instructions – which must be adequately recorded and made available for examination by the customer. They must leave no room for error, but equally they should be flexible and written in such a way that it is possible to modify their content to reflect changing circumstances.

FIG. 2.42 The lack of a Quality Plan can have disastrous results!

A well-thought-out Quality Plan will divide the project into stages, show what type of inspection has to be completed at the beginning, during or end of each stage and indicate how these details should be recorded. The Quality Plan should be planned and developed in conjunction with design, development, manu-facturing, subcontracted pre- and post-installation work and ensure that all functions have been fully catered for.

FIG. 2.43 What should be covered by a Quality Plan

2.6.1 What should be covered by a Quality Plan?

One of the main objectives of quality planning is to identify any special or unusual requirements, processes and techniques (including those requirements that are unusual by reason of newness, unfamiliarity, lack of experience and/or absence of precedents). As pointed out in ISO 9000, if the contract specifies that a Quality Plan is required, then that Quality Plan should fully cover the areas listed in Fig. 2.43.

2.6.2 Management responsibility

The Quality Plan should show who is responsible for:

- ensuring that activities are planned, implemented, controlled and monitored;
- communicating requirements and resolving problems;
- reviewing audit results;
- authorising exemption requests;
- implementing corrective action requests.

Where the necessary documentation is already available under an existing QMS, the Quality Plan need only refer to a specific situation or specification.

2.6.3 Contract review

Contract review should cover:

- when, how and by whom the review is made;
- how the results are to be documented;
- how conflicting instructions or ambiguities are resolved.

2.6.4 Design control

Design control should indicate:

- when, how and by whom the design process, validation and verification of the design output is carried out, controlled and documented;
- any customer involvement;
- applicable codes of practice, standards, specifications and regulatory requirements.

2.6.5 Documentation

Document and data control should refer to:

- what is provided and how it is controlled;
- how related documents will be identified;
- how and by whom access to the documents can be obtained;
- how and by whom the original documents are reviewed and approved.

2.6.6 Purchasing

Under the heading of purchasing, the following should be indicated:

- the important products and services to be purchased;
- the source and requirements relating to these;
- the method, evaluation, selection and control of subcontractors;
- the need for a subcontractor's Quality Plan in order to satisfy the regulatory requirements applicable to the purchase of products and services.

2.6.7 Customer-supplied products and services

Customer-supplied products and services should refer to:

- how they are identified and controlled;
- how they are verified as meeting specified requirements;
- how nonconformance is dealt with.

2.6.8 Product and service identification and traceability

If traceability is a requirement, then the plan should:

- define its scope and extent (including how products and services are identified);
- indicate how contractual and regulatory authority traceability requirements are identified and incorporated into working documents;
- indicate how records are to be generated, controlled and distributed.

2.6.9 Process control

Process control may include:

- the contract-specific QPs and WIs;
- process steps;
- methods to monitor and control processes;
- service/services characteristics;
- reference criteria for workmanship;
- special and qualified processes;
- tools, techniques and methods to be used.

2.6.10 Inspection and testing

Inspection and testing should indicate:

- any inspection and test plan;
- how products and services shall be verified;
- the location of inspection and test points;
- procedures and acceptance criteria;
- witness verification points (customers as well as regulatory);
- where, when and how the customer requires third parties to perform:
  - type tests;
  - witness;
  - service/product verification;
  - material, service, product, process or personnel certification.

2.6.11 Inspection, measuring and test equipment

Inspection, measuring and test equipment should:

- refer to the identity of the equipment;
- refer to the method of calibration;
- indicate and record calibration status and usage of the equipment;
- indicate specific requirements for the identification of inspection and test status.

2.6.12 Nonconforming products and services

Under the heading of nonconforming products and services, an indication should be given of:

- how such products and services are identified and segregated;
- the degree or type of rework allowed;
- the circumstances under which the supplier can request concessions.

Details should also be provided with respect to:

- corrective and preventive action;
- handling, storage, packaging, preservation and delivery.

2.6.13 Other considerations

Quality Plans should:

- indicate key quality records (i.e. what they are, how long they should be kept, where and by whom);
- suggest how legal or regulatory requirements are to be satisfied;
- specify the form in which records should be kept (e.g. hardcopy, flash drives, disk, cloud, etc.);
- define liability, storage, retrievability, disposition and confidentiality requirements;
- include the nature and extent of quality audits to be undertaken;
- indicate how the audit results are to be used to correct and prevent recurrence of deficiencies;
- show how the training of Staff in new or revised operating methods is to be completed.

Where servicing is a specified requirement, suppliers should state their intentions to assure conformance to applicable servicing requirements, such as:

- regulatory and legislative requirements;
- industry codes and practices;
- service level agreements;
- training of customer personnel;
- availability of initial and ongoing support during the agreed time period;
- statistical techniques, where relevant.

Note: ISO 10005 (*Quality Management Systems – Guidelines for Quality Plans*) provides useful guidance on how to produce Quality Plans, as well as including helpful suggestions on how to maintain an organisation's quality activities.

2.7 WHAT IS A QUALITY PROCEDURE?

Quality Procedures are used to implement the quality processes of an organisation. QPs detail *what* has to be carried out to meet the requirements of these primary and secondary processes and their associated Quality Policies. Without procedures, an organisation's best intentions will not always be met.

FIG. 2.44 Written procedures make all the difference!

Think of QPs as clear, concise instructions. For example, management decrees that all problems found within the organisation must go through a problem-solving process (i.e. management sets a policy). A member of Staff couldn't be expected to know how to do this without clear instructions. Even worse, the entire work force would have their own ideas about solving problems and further problems would arise because of this.

It is, therefore, essential that *all* QPs are written down so that everyone knows what to do.

 But beware of falling into the trap of writing a book for each procedure, as human nature puts us off reading masses of text! It is *not* mandatory to write procedures for everything.

2.7.1 What is the best way to write a Quality Procedure?

A simple rule is don't bother unless you really have to. A degree of common sense must prevail when writing procedures, and you should always consider the known competency and knowledge of the personnel performing them.

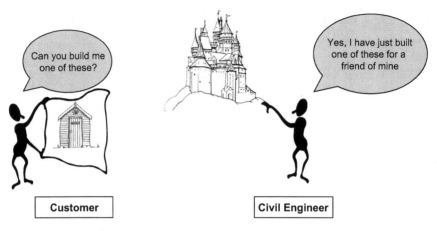

FIG. 2.45 Can you design this?

If you employ highly skilled professionals, it would be foolish to spend hours writing a procedure on something which they are already qualified to do or for which standards and guidelines already exist. For example, you would not tell a civil engineer how to design a beach hut.

On the other hand, you will probably want to write a procedure where you use untrained Staff to perform it. A simple rule is: don't bother unless you really have to. A degree of common sense must prevail when writing procedures, and due consideration must be given to the competency of the personnel performing them.

The example on page 55 shows part of a typical procedure for resolving issues (such as some of the problems involved with providing a service).

This easy-to-follow flowchart can be enhanced with explanatory text so that the entire process is clearly understood and easily trained. The benefit of flowcharts is that they can be stuck up on notice boards so that the people who have to implement the procedure can readily see what they are expected to do.

 The old adage that a picture paints a thousand words stands true for QPs. No one likes reading, and visually representing an organisation's QPs has always proved successful! If you can do *without* a written procedure, do so and use a simple process map instead.

2.7.2 What should go into a Quality Procedure?

QPs form the basic documentation used for planning and controlling all activities that impact on quality. They detail how an organisation's Core Policy is to be implemented by adding the meat to the process-specific policies. They should cover all the applicable elements of ISO 9001:2015 and detail procedures that concern the organisation's actual method of operation. These will normally remain relatively constant, regardless of the product, services, system or process being supplied.

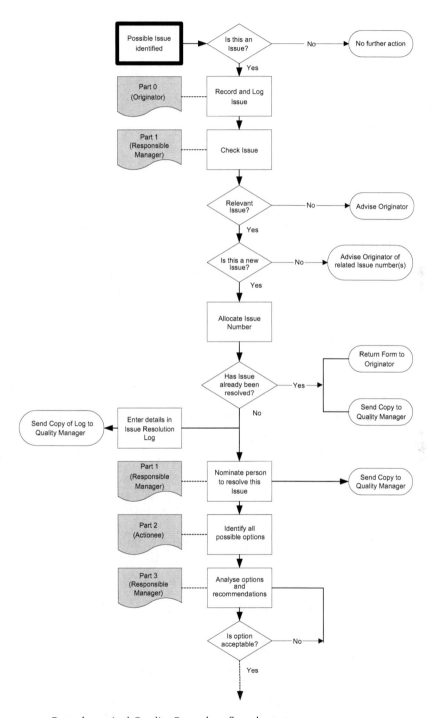

FIG. 2.46 Part of a typical Quality Procedure flowchart

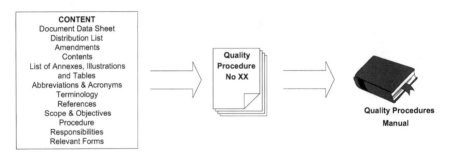

FIG. 2.47 What goes into a Quality Procedure?

Each QP should cover a specific part of the Core Business Process or one of its supporting processes, e.g. contract review, document control, audit procedures or training, and should be easily traced back to the process-specific policies dictated by Top Management.

QPs should not normally include technical requirements or specialist procedures required for the manufacture of a product or delivery of a system/service. These sorts of details are generally explained in WIs (see Chapter 2.8 for further details). QPs can (and usually do) form a large bulk of the QMS.

The layout and format of QPs should be consistent so that Staff can become accustomed to a familiar structure. This also helps to ensure systematic compliance with the ISO 9001:2000 standard.

QPs will mainly include (but are not limited to) the following:

- **Document data sheet**: all the salient information about the document – file name, who wrote it, a summary of the contents, when it was approved, who approved it, etc.;
- **Distribution list:** a record of everyone who has a controlled copy of the document;
- **Amendments:** a record of all changes made to the document;
- **Contents list:** a list of all the chapters, sections, parts and annexes, etc. that make up this document;
- **List of annexes:** all parts of a document should be traceable, especially when they are in separate volumes;
- **List of illustrations/tables:** a list of all the figures and tables included in the document;
- **Abbreviations and acronyms:** an explanation of any abbreviations or acronyms used in the document;
- **Terminology:** an explanation of any technical or confusing terminology used in the document;
- **References:** any reference material that is specifically referred to in the document;
- **Scope and objectives:** this should list why you need the procedure, what it is for, the area covered and any exclusions;

- **Procedure overview and procedure**: this is the main part of the document and details in clear, concise and unambiguous terms the actions and methods to be used. Ideally the procedure should be detailed in a logical order with the aid of flowcharts;
- **Responsibilities**: clear specifications of who is responsible for implementing the procedure and who can carry it out including (if necessary) minimum training requirements;
- **Relevant forms**: the identification of any forms, paperwork or computerised forms required to implement the procedure.

The contents of a typical Quality Procedure for the control of documented information are shown below.

QP/1: Control of documented information

1   Administration of documented information
     1.1   Scope
     1.2   Responsibility
     1.3   Definition
     1.4   Document administration number
     1.5   File numbering system
     1.6   Documents produced by LUR
         1.6.1   File reference
         1.6.2   Version numbering
     1.7   Letters produced by LUR
     1.8   Faxes produced by LUR
     1.9   E-mails produced by LUR
     1.10   Contracts and minutes produced by LUR
     1.11   Documents received by LUR
     1.12   E-mails received by LUR personnel
     1.13   Storing files on the server
     1.14   Filing of documents (hard copies)
     1.15   Old and obsolete documents
     1.16   Controlled documents
     1.17   Headed paper
     1.18   Document distribution
     1.19   Draft documents
     1.20   Approved documents
     1.21   Internal distribution
     1.22   External distribution
     1.23   Press notices etc.
     1.24   Software
         1.24.1   Word processing
         1.24.2   Spreadsheets and graphics

2.8 WHAT IS A WORK INSTRUCTION?

Author's Hint

Although ISO 9001:2015 states that Work Instructions (WIs) are no longer required (they now refer to them as '*documented information that defines the activities to be performed and the results achieved*'!), it seems nowadays (well it certainly does to me!) that whenever you purchase something that you will need to assemble yourself (e.g. a fruit cage) or use (e.g. a simple car hoover) a WI is included in the packing box. So whatever you call this – a WI or 'written guidance' – it is still part of the overall Quality Assurance chain – in my opinion.

In a nutshell:

- Procedures: tell you who does what and when;
- Work Instructions: tell you *how* to do something at a more granular level.

Work Instructions provide the 'nitty gritty' detail required to carry out a specific job in an exact manner and to a predetermined standard. They detail how an organisation manufactures a product or supplies a process or service – as well as the controls that it has in place to ensure the quality of that product and service, etc. is consistent.

FIG. 2.48 Written instructions should leave no room for error

WIs describe, in detail, procedures such as 'what is to be done', 'who should do it', 'when it should be done', 'what supplies, services and equipment are to be used' and 'what criteria have to be satisfied'. These WIs should be regularly reviewed for their continuing acceptability, validity and effectiveness.

Inferior or poor design, ambiguous specifications, incomplete or inaccurate WIs and methods, nonconformance etc. are the most frequent causes of defects during manufacture or the delivery of a process or service. In order that management can be sure that everything is being carried out under the strictest of controlled conditions, it is crucial that all WIs (in fact any written instruction) are clear, accurate and fully documented.

Good WIs avoid confusion and show exactly what work has to be done or what services are to be provided. They also delegate authority and responsibility.

Without a written guide, differences in policies and procedures can easily arise and these variations can result in confusion and uncertainty.

As ISO reminds us: *'Instructions provide direction to various ISO 9000 levels of personnel. They also provide criteria for assessing the effectiveness of control and the quality of the material, ensure uniformity of understanding, performance and continuity when personnel changes occur. They provide the basis for control, evaluation and review'.*

2.8.1 What should go into a Work Instruction?

FIG. 2.49 What goes into a Work Instruction?

In summary, ideally a WI should contain:

- **Document data sheet**: all the salient information about the document – file name, who wrote it, a summary of the contents, when it was approved, who approved it and a record of all changes made to the document etc.;
- **Contents list:** a list of all the chapters, sections, parts and annexes, etc. making up this document;
- **List of annexes:** all parts of a document should be traceable, especially when they are in separate volumes;
- **References:** a clear reference to any material or technology that is specifically referred to in the document. Also any associated QPs, etc.;
- **Scope and objectives:** this should define exactly what the WI is needed for. Normally this is a very simple statement because a WI would usually be limited to one process (e.g. the WI illustrated in Fig. 2.48 detailing the actions to be taken to dig a square hole);
- **Procedure:** again, referring to Fig. 2.48 as an example, this will state the manner of production, installation or application where the absence of such controls would adversely affect the quality of the hole to be dug using hand shovels only, and would require temporary shoring to prevent collapse. Consideration should also be given to any safety implications that may exist when carrying out the process;
- **Responsibility:** the WI must clearly state who can carry out the process;
- **Relevant forms:** the identification of forms – hardcopy or computerised – required to implement the WI.

2.8.2 How many Work Instructions can I have?

The manufacture of a device or the delivery of a process or service may require the completion of more than one WI. It is perfectly acceptable, indeed desirable, to separate processes into a number of WIs because:

- it would be very difficult to write a single WI for large items, such as building an aircraft or laying on the catering services for the Royal Tournament;
- each WI may require Staff with different levels of training and qualifications;
- a particular contract may only require the completion of certain WIs;
- small, concise WIs are more easily revised.

2.9 WHO CONTROLS QUALITY IN AN ORGANISATION?

Quality Management includes all the activities that organisations use to direct, control and coordinate quality. These activities include formulating a Quality Policy, setting quality objectives and establishing quality processes. They also include quality planning, Quality Control, Quality Assurance and quality improvement.

2.9.1 Who has overall responsibility?

Previously the overall responsibility for an organisation's Quality Management System was the sole responsibility of the Quality Manager, who was (i.e. in ISO 9001:2008) often referred to as the 'Management Representative'. In the new

Who Controls Quality in a company?

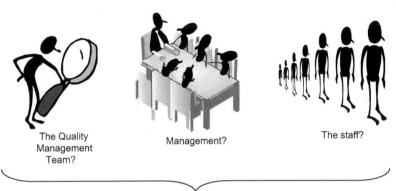

The Quality Management Team?   Management?   The staff?

FIG. 2.50 Everyone has a control on quality

ISO 9001:2015, the title Quality Manager has been dropped and ownership of the QMS does not centre on a single individual; in future, these duties may be assigned to any role or be split across several roles.

There have been numerous thoughts and discussions about ISO's new 'ruling' within professional bodies and on social media, and the general feeling appears to be that as this is *not* conceived to be a mandatory requirement, an organisation may, if they so wish, continue to use the term Quality Manager, *but* with the proviso that Top Management will now have to play a far more important role in overseeing management within their organisation than before – and: *Everyone within an Organisation Has a Part to Play in Quality*!

So what responsibilities do each of these groups of people have?

2.9.2 Management responsibilities

The main requirements of an organisation's management are that they:

- fully endorse and are committed to the development and implementation of their organisation's quality system;
- develop and establish a Quality Policy that can be supported by measurable quality objectives;
- plan, develop, implement, improve and modify their Quality Management System;
- define and communicate organisational responsibilities and authorities;
- ensure that everyone within the organisation plays a role in being part of the effective implementation of the Quality Management System;
- establish (and use) internal communication processes;
- review the Quality Management System at planned intervals.

Having established their overall position, the management will then have to:

- ensure that the organisation's QMS *always* meets the requirements of the national, European or international standard to which the particular organisation has chosen to work and, where this fails to happen, to see that corrective actions are carried out;
- define objectives such as fitness for use;
- ensure that the performance, safety and reliability of their products and services are correct and that costs associated with these objectives are kept to a reasonable level.

2.9.3 The Quality Management Team

It is quite normal (especially for large organisations) that a completely separate and independent division, headed up by someone from Top Management (usually the Quality Manager) who deals solely with quality matters. The organisation of this section 'could' look something like that shown in Fig. 2.51.

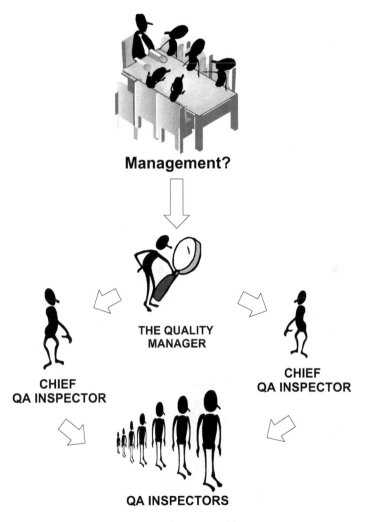

Management?

**THE QUALITY
MANAGER**

**CHIEF
QA INSPECTOR**

**CHIEF
QA INSPECTOR**

QA INSPECTORS

FIG. 2.51 Quality management structure for a typical large organisation

For organisations that cannot justify the cost of employing full-time inspectors, other options are available, such as:

- selecting personnel from existing Staff who are not directly involved with a production process. They are then able to act as independent unbiased assessors;
- employing third party quality consultants, on a temporary basis, to carry out fully independent Quality Controls.

The Quality Team members (i.e. 'QA Inspectors') working under the Quality Manager are part of an organisation adjudged competent to carry out QA duties.

2.9.4 The Staff

Your Staff are at the sharp end of delivering quality. They are responsible for implementing the QC processes (i.e. the QPs and WIs described in chapters 2.7 and 2.8) that will ensure the desired level of quality is consistently applied to the product. Of course they are not responsible for setting the level of quality but, so long as they have been clearly briefed on what is required and have received the appropriate training to do the job, they will be capable of delivering that level of quality.

It is vital that a workforce is as committed to quality as the management. A committed workforce will look after your organisation. A workforce who is empowered to implement quality (and are allowed to have an input in defining and improving it) will be highly motivated. Morale will improve as Staff will feel that they are doing a good job that they can be proud of.

In summary, an organisation workforce has the responsibility for:

- working in accordance with the predefined QPs and WIs;
- refusing to accept anything that is substandard;
- having an active role in quality improvement;
- having an input into defining levels of quality (after all, they know better than anybody what can be achieved);
- delivering the level of quality specified in the QMS.

In short, your Staff is your greatest asset.

FIG. 2.52 Recognise your greatest asset!

2.10 HOW QUALITY HELPS DURING THE LIFE CYCLE OF PRODUCTS AND SERVICES

As described in Chapter 2.5, a process is 'a system of activities, which uses resources to transform inputs into outputs'.

For organisations to function effectively they will not only have to identify and manage a number of interlinked processes, they will also have to incorporate QA into their management system and the life cycle of their products or services.

This 'life cycle' basically consists of five separate stages, namely:

- the design stage;
- the manufacture of a product or the implementation of a service;
- the acceptance stage;
- the in-service stage;
- the end-of-life stage.

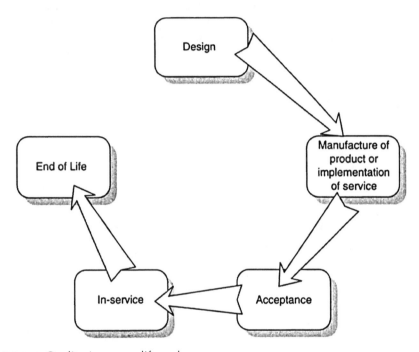

FIG. 2.53 Quality Assurance life cycle

Author's Hint

Although some of the stages of the products' or services' life cycle may not be necessary for your particular organisation – e.g. if you are only responsible for the installation and maintenance of a predesigned and manufactured product as opposed to making the thing yourself – the description of these stages could be a useful reminder if the product you are installing or service you are providing proves faulty owing to a manufacturing or design fault and you need to investigate that fault.

2.10.1 Design stage

'Quality must be designed into a product before manufacture or assembly'
(ISO 9000)

Throughout the design stage of a product or service, the quality of that design must be regularly checked. Quality Procedures (QPs) have to be planned, written and implemented so as to predict and evaluate the fundamental and intrinsic reliability of the proposed design.

It is at this time that the proposed design is verified against the customer's requirements to ensure it will actually deliver the intended functionality.

Throughout the design stage the quality of that design must be regularly checked. QPs have to be planned, written and implemented so as to predict and evaluate the fundamental and intrinsic reliability of the proposed design.

It is at this time that the proposed design is verified against the customer requirements to ensure it will actually deliver the intended functionality.

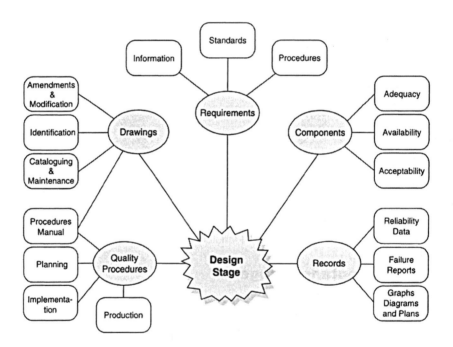

FIG. 2.54 Design stage

2.10.2 Manufacturing stage

'Manufacturing operations must be carried out under controlled conditions'
(ISO 9000).

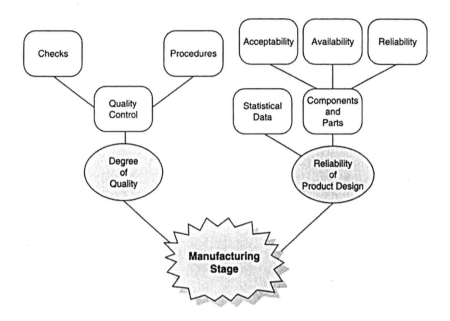

FIG. 2.55 Manufacturing stage

During all manufacturing or production processes (and throughout early in-service life), products and services must be subjected to a variety of Quality Control procedures and checks in order to evaluate the degree of quality. These controls will ensure the products and services comply with your predetermined documented requirements.

2.10.3 Acceptance stage

'The Quality of a product must be proved before being accepted' (ISO 9000).

During the acceptance stage, the product and/or service is subjected to a series of tests designed to confirm that the workmanship fully meets the levels of quality required (or stipulated) by the user and that the product or service performs the required function correctly. Tests will range from environmental tests of individual components to field testing of complete products.

This acceptance stage is generally termed the 'validation' phase, where finished products and services are checked to ensure that they comply with the original requirements.

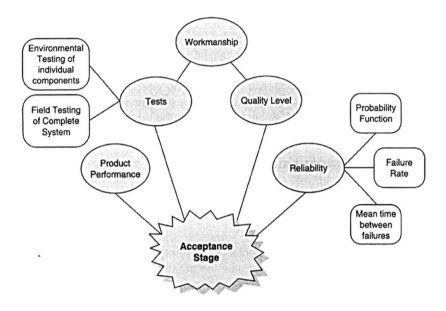

FIG. 2.56 Acceptance stage

2.10.4 In-service stage

'*Evaluation of product performance during typical operating conditions and feed-back of information gained through field use – improves product capability*' (ISO 9000).

During the in-service stage the equipment or service user is, of course, principally concerned with reliability.

2.10.5 End-of-life stage

Designing, manufacturing, accepting and using products and services is not the full story. Eventually they will come to the end of their useful life either through age, fault or, more than likely (because it has been designed and/or built to such a high degree of quality!), it has been overtaken by technology. Before throwing the redundant piece of equipment onto the rubbish heap, however, it is essential that a fully documented historical record, of its design, use, problems, advantages and disadvantages, etc. is assembled.

So, if you can incorporate quality into all five stages within the life of products and services, *then* you can have total control over your own success . . . or failure!

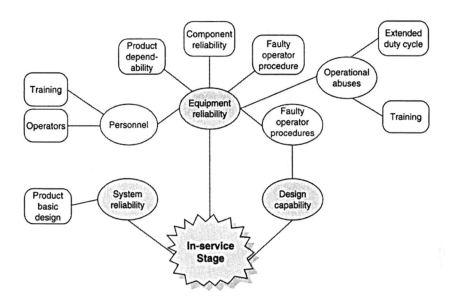

FIG. 2.57 In-service stage

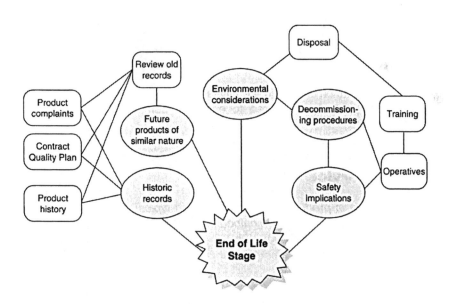

FIG. 2.58 End-of-life stage

2.11 WHAT ARE THE PURCHASER'S RESPONSIBILITIES?

As previously described, Quality Assurance is concerned with consistency of quality and an agreed level of quality. To achieve these aims an organisation must be firmly committed to the fundamental principle of always supplying the right quality of product or service. Equally, a purchaser must be committed to the fundamental principle of only accepting the right quality product – especially when quite a number of problems associated with a product's quality are usually the fault of the purchaser!

Obviously the purchaser can only expect to get what he ordered. It is, therefore, extremely important that the actual order is not only correct, but also that the purchaser provides the product or service provider with all the relevant (and accurate) information required to complete the task.

FIG. 2.59 Insufficient information from the purchaser!

There is little point in trying to blame the product or service provider when the finished deliverable doesn't come up to expectation because of an unsatisfactory design provided by the purchaser. In certain cases, (for example when the requirements of the item cannot easily be described in words), it could be very helpful if the purchaser was to provide a drawing as a form of graphic order (i.e. a WI!). In such cases, this drawing should contain all the relevant details such as type of material to be used, the material grade or condition, the specifications that are to be followed and, where possible, the graphic order/drawing should be to scale.

If this approach proves impractical, then the order would have to include all the relevant dimensional data, sizes, tolerances, etc., or refer to one of the accepted standards.

Having said all that, it must be appreciated that the actual specification being used is also very important for it sets the level of quality required and, therefore, directly affects the price of the article. Clearly, if specifications are too demanding

FIG. 2.60 A good specification provided by the purchaser

then the final cost of the article will be too high. If specifications are too vague or obscure, then the supplier will have difficulty in assembling, or even designing, the object or may even be unable to get it to work correctly.

The choice of product or service provider is equally important. It is an unfortunate fact of life that purchasers usually consider that the price of the article is the prime, and in some cases, even the *only* consideration. Buying cheaply is obviously *not* the answer because if a purchaser accepts the lowest offer, all too often they will find that delivery times are lengthened (because the manufacturer/supplier can make more profit on other orders), the article produced does not satisfy their requirements and, worst of all, the quality of the article is so poor that they have to replace the device well before its anticipated life cycle has been completed.

If a product or service provider has received official recognition that the quality of his work is up to a particular standard, then the purchaser has a reasonable guarantee that the article being produced will be of a reasonable quality – always assuming that the initial order was correct! Official recognition is taken to mean that an organisation has been assessed and certified to a recognised quality standard such as ISO 9001. In other words the level of quality can be *proved*.

2.12 WHAT ARE THE SUPPLIER'S RESPONSIBILITIES?

The term 'supplier' relates to organisations that produce products *or* provide services. The supplier's prime responsibility must always be to ensure that anything *and* everything leaving their organisation conforms to the specific requirements of the purchaser – particularly with regard to its quality.

The simplest way of doing this is for the supplier to ensure that their particular office, production facility or service outlet fully complies with the requirements of the quality standards adopted by the country to which they are providing this service and the country to which they intend supplying the component, equipment or system.

To do this they must of course first be aware of the standards applicable to that country, know how to obtain copies of their standards, how to adapt them to their own particular environment and how to get them accepted by the relevant authorities.

Every country has its own set of recognised quality management standards by which suppliers can be assessed and certified. The following are the most frequently used certification and guideline standards used within the UK.

- **BS EN ISO 9001:2015**: Quality Management Systems. Requirements;
- **BS EN ISO 14001:2015**: Environmental management systems. Requirements with guidance for use;
- **BS OHSAS 18001:2007**: Occupational health and safety management systems. Requirements.

Although an organisation can set out to abide by accepted standards, unless it is able to achieve this aim it will fail in its attempt to become a recognised supplier of quality goods. The main points it should note are that:

- all managerial Staff, from the most junior to the most senior, must firmly believe in the importance of Quality Control and Quality Assurance and understand how to implement them;
- managerial Staff *must* create an atmosphere in which Quality Assurance rules are obeyed and not simply avoided just because they are inconvenient, time consuming, laborious or just too boring to bother with;
- there has to be an accepted training scheme to ensure that all members of the organisation are regularly brought up to date with the ongoing and latest requirements of Quality Assurance;
- management must ensure that everyone in their organisation accepts that they are **ALL** responsible for the quality of the products and services provided by their company.

FIG. 2.61 If you've got it, flaunt it!

In addition, the supplier will have to provide proof that they are supplying a quality product. This is actually a 'measurement of their Quality Control' and usually takes the form of a supplier's evaluation, surveillance and audit. The evaluation is carried out by either (1) the prospective purchaser of the product or (2) an accredited body (such as Lloyds, BSI, SGS or ISOQAR) and which, if successful, will allow the supplier to proudly display a compliance certificate and to use the recognised 'quality mark' on its stationery and marketing literature.

2.13 QUALITY MANAGEMENT AND COMPUTER TECHNOLOGY

Nowadays, computer technology appears in just about every walk of life and in most instances provides great benefits. Quality Management is no exception to this rule, as computers offer significant advantages when implementing and controlling a Quality Management system.

One of the most onerous tasks of Top Management is to ensure that the efficacy of their Quality Management System is maintained. Now if they only have one item of documented information, this is not much of a problem. Difficulties arise when there are numerous items of documented information dotted around the business, all of which need to be controlled to ensure their content is current. This problem is further compounded when a business is spread throughout a country or, in the case of multi-nationals, throughout the world.

The same problems arise when you need a number of isolated personnel to perform the same process. It is very difficult to ensure everyone has the same version to work from.

FIG. 2.62 Document control gone crazy

This is where computer technology comes into its own. Through the use of a company intranet it is possible for the Top Management to:

- maintain one centrally located management system;
- permit all personnel access to the same information;
- control only one set of documented information.

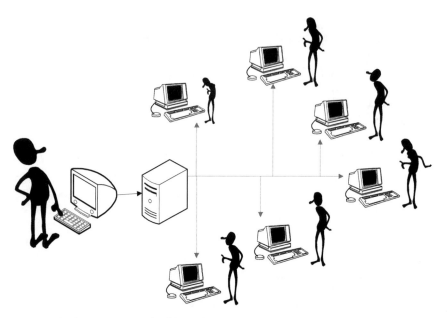

FIG. 2.63 An intranet can make life easier

It is clear that such a system is very efficient and hence cost effective. So what is a company intranet? Put crudely, it is like a mini-version of the World Wide Web, except that only *your* Staff can access the website. So, by turning the contents of your documented information onto an intranet website, all your personnel can access the current version.

Computers are therefore ideal as a means of control and distribution of processes, procedures and WIs. They are also an ideal means of producing and then seamlessly combining your process maps. There are any number of software packages offering process-mapping capabilities. Indeed, Microsoft's own Visio® package is more than adequate for most purposes, although you can purchase bespoke software should you have more specific needs.

By using process-mapping software it is possible to set up standard templates to ensure your maps all have a consistent appearance and convention. This is of importance if you are getting a number of personnel to draw up processes.

You will recall from the chapter on Quality Processes (see Chapter 2.5) that these are often complex and require a number of levels to fully detail the activities. This can result in problems finding the way between the maps, especially when they are in paper form. By converting your process maps into web pages it is possible to include electronic links. This is called 'hyperlinking' and enables separate processes to be connected and hence aids navigation between them.

In the following diagram (Fig. 2.64) you will see, by attaching hyperlinks between points on different process maps, it is possible to jump between them and consequently maintain continuity and the sequence of processes.

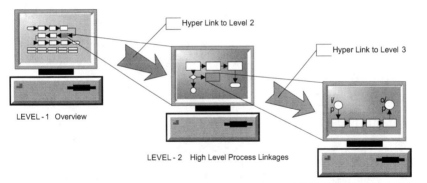

FIG. 2.64 Hyperlinked process maps

One other area where computers come into their own is in communication. ISO 9001:2015 requires Top Management to ensure that appropriate communication processes are established within the organisation and that communication takes place regarding the effectiveness of the QMS. What better way than using your office e-mail facility.

Consider also the possibility that you may need to have current versions of legislation made available to your personnel. Yet again the centralisation of this documentation within an intranet will avoid unnecessary duplication.

In summary, central management of the master copy of all your organisation's documented information on computer will greatly assist quality management for:

- document control;
- document distribution;
- process map development;
- process map navigation through hyperlinking;
- communicating quality issues;
- immediate availability of applicable specifications, standards and legislation.

FIG. 2.65 A computer is a real friend of quality!

In other words, computers make you more efficient and efficiency means lower costs.

When developing your QMS, serious consideration should be given to the use of computer technology. The days of paper-based systems are numbered. Computers will make your life a whole lot easier.

Author's Note

Having now covered all the bits and pieces that make up a Quality Management System, the responsibilities for quality from the organisation, purchasers and suppliers, you might well be thinking 'when did all this quality thing start?' And so, in the next Chapter, I shall attempt to provide you with a road map of where ISO 9001 came from, its current position as a management system and what the future holds for this particular international standard.

The history of quality standards

CONTENTS

Author's Note

The history of quality management can be traced way back to medieval Europe where work completed by journeymen and apprentices was evaluated an inspected by a skilled worker to ensure that the quality of the finished product was up to the required standard, and most importantly, ensured the satisfaction of the buyer. Although quality management had undergone a number of important changes since that time, the end goal has always been the same, but it wasn't until the 1920s that quality management systems, as we know them today, started to surface and the following chapter is intended to provide you with a sort of ISO 9001:2015 'time map'.

The quality movement can trace its roots back to 13th century medieval Europe, where craftsmen began organising themselves into unions called guilds. These guilds were responsible for developing strict rules for product quality, and inspection committees enforced the rules by marking flawless goods with a special mark or symbol. Craftsmen themselves often placed a second mark on the goods they produced. At first this mark was used to track the origin of faulty items, but over

FIG. 3.1 Typical silvermakers' mark

(Courtesy of Stingray Quality Management Consultants®)

time the mark came to represent a craftsman's good reputation and symbolised each guild member's obligation to satisfy his customers and enhance the trade's reputation.

Until the early 19th century, manufacturing in the industrialised world tended to follow this craftsmanship model until the Industrial Revolution and the growth of the factory system – with its emphasis on product inspection – became increasingly important, and gradually manufacturers began to include quality processes in their quality practices.

In the 1920s, a Munitions Standard was developed by the UK Ordnance Board to guarantee that bullets used during the 1st World War were good (and safe!) enough to be fired. This standard is now known as Def Stan 13–131 and is the benchmark by which all munitions are currently measured.

FIG. 3.2 This is *not* a good check for quality!

Quite a lot of people have said that today's ISO 9000 originated from this 1920s munitions standard. I tend to believe, however, that the actual 'start' of ISO 9000 was probably during the US Navy Polaris submarine programme in the late 1950s when Admiral Hyman G. Rickover – who was the head of the US Nuclear Navy, and renowned for having a ruthless disposition and a very hot temper – became frustrated with the number of Polaris submarines that were unseaworthy and

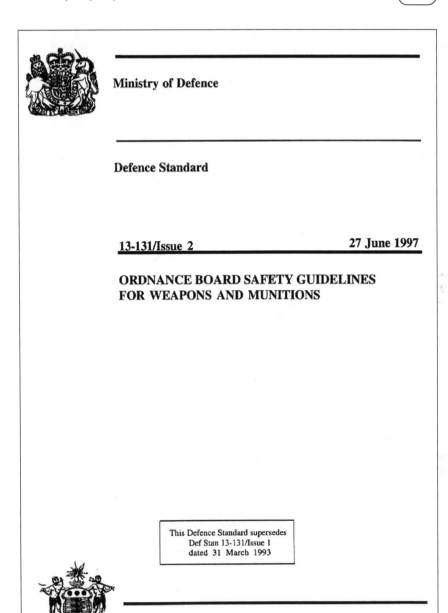

FIG. 3.3 Def Stan 13–131 (updated version of previous 1920 Munitions Standard)

moored up at dockyards mainly due to defects, errors and general quality break-downs.

He took 30 fresh graduates from Harvard, gave each of them a list of subcontractors to visit and investigate, a time scale and report format and then sent them out into the industrial jungle!

When they returned and the results were analysed, Rickover discovered that there were 18 major and semi-major items which were the most common, or root cause, to all of the problems being experienced. For example:

FIG. 3.4
Admiral Hyman G. Rickover – the founding father of quality?

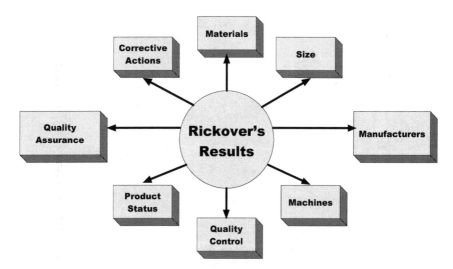

FIG. 3.5 Admiral Rickover's major survey findings

As shown in Fig 3.5 and Table 3.1, the most important failures which needed to be immediately addressed were:

- **wrong materials obtained:** failure to specify totally and exactly what was required on the purchase orders;
- **items made to wrong dimensions**: failure to withdraw obsolete drawings when essential design changes had been made;
- **items from different manufacturers would not fit together at the quayside:** failure to calibrate measuring instruments to reliable reference standards, which differed between organisations and districts;

TABLE 3.1 Admiral Hyman G. Rickover's findings

Wrong materials	Wrong materials obtained
Wrong dimensions	Items made to wrong dimensions
Different manufacturers	Items from different manufacturers would not fit together at the quayside
Inefficient machines	Machines operated incorrectly – defects made in products through lack of skill
Quality Control	Lack of Quality Control and defective items dispatched
Product status	Status of products unknown
Quality Assurance	Products in doubt and no corrective actions taken
Corrective action	Corrective actions not recognised or at best delayed, etc.

- **machines operated incorrectly:** defects made in products through lack of skill – failure to train operators in performing vital tasks;
- **lack of Quality Control and defective items dispatched:** failure to inspect effectively;
- **status of products unknown:** no records available to show what had or had not been done;
- **lack of Quality Assurance:** no appointed person to ensure operations were conducted properly;
- **corrective actions not recognised or at best delayed, etc.:** Top Management unaware of what was and was not happening.

A total of 18 points emerged from the survey and these were then used as the cornerstone for quality in the American Space Research Programme (i.e. by NASA) and eventually became the basis of the first NATO Allied Quality Assurance Publications (AQAP) specifications, which defined the QMS requirements to be adopted by all military subcontractors for all defence products.

On a similar note to the Polaris problems, NASA's first major interplanetary project – the Ranger probe, designed to impact on the surface of the Moon – nearly failed, due to a total lack of project management techniques.

3.1 1979 – BS 5750:1979 PARTS 1, 2 AND 3

The year 1979 was an important one for quality standards within the UK. The British Standards Institution (BSI) had already published a number of guides on QA (e.g. BS 4891:1972 *A Guide to Quality Assurance*), but with increased requirements for some sort of auditable Quality Assurance, BSI set up a study group to produce an acceptable document that would cover all requirements for a two-party manufacturing and/or supply contract.

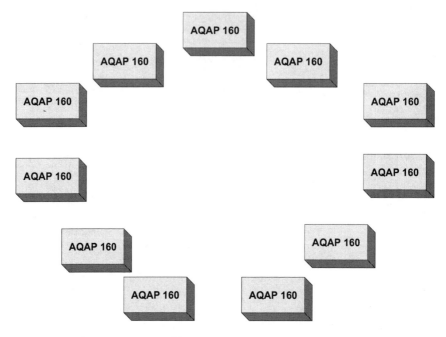

FIG. 3.6 List of current AQAP publications

TABLE 3.2 List of current AQAP publications

AQAP 160	NATO Integrated Quality Requirements for Software throughout the Life Cycle
AQAP 2000	NATO Policy on an Integrated Systems Approach to Quality through the Life Cycle
AQAP 2050	NATO Project Assessment Model
AQAP 2070	NATO Mutual Government Quality Assurance (GQA) Process
AQAP 2105	NATO Requirements for Deliverable Quality Plans
AQAP 2110	NATO Quality Assurance Requirements for Design, Development and Production
AQAP 2120	NATO Quality Assurance Requirements for Production
AQAP 2130	NATO Quality Assurance Requirements for Inspection and Test
AQAP 2131	NATO Quality Assurance Requirements for Final Inspection
AQAP 2210	NATO Supplementary Software Quality Assurance Requirements to AQAP 2110
AQAP 2310	NATO Quality Management System Requirements for Aviation, Space and Defence Suppliers

FIG. 3.7 The main parties involved in the transition to ISO 9000

Reproduction of these logos courtesy of the International Organization for Standardization, British Standards Institution and the American National Standards Institute

This became the BS 5750 series of standards on quality systems, which were first published in the UK during 1979.

These standards supplied guidelines for internal QM as well as external QA. They were quickly accepted by manufacturers, suppliers and purchasers as being a reasonable minimum level of Quality Assurance to which they could be expected to work. The BS 5750 series thus became the 'cornerstone' for national quality.

But in the meantime, the USA had been working on their American National Standards Institute (ANSI) 90 series, and other European countries were also busily developing their own sets of standards. Quite naturally, however, as BSI had already produced and published an acceptable standard, most of these national standards were broadly based on BS 5750.

The concept was further developed by the defence, power generation, automobile and textile industries and gradually expanded from 18 initial points to 20 basic elements. These were:

- **Management responsibility**: Management should define and document a Quality Policy, an organisation structure, including responsibility and authority. Management should make available verification resources (inspectors), appoint a management representative and carry out management reviews.
- **Quality system**: The quality system must be documented, including a manual, procedures and Work Instructions.
- **Contract review**: A procedure for performing contract review – documenting what was agreed with the customer – should be written stating clearly the criteria for contractual obligations to be met.

- **Design control**: Procedures should define how the organisation designs its product, and controls any design changes.
- **Document control**: Procedures and Work Instructions must be approved before issue and on subsequent changes. Control of documents should encompass availability, distribution, issue level, revision and obsolescence.
- **Purchasing**: Suppliers should be assessed and monitored, incoming products should be verified.
- **Customer-supplied stock**: Customer-supplied stock should be subject to procedures for identification, inspection, storage and periodic maintenance. There should also be a procedure for reporting and recording lost or damaged stock.
- **Product identification and traceability:** A company-wide procedure should detail how items and equipment are to be identified at all stages, from receipt to despatch. Where traceability is required, a unique identification should be used and recorded.
- **Process control**: Work Instructions defining when, how and what should be done need to be documented and made available.
- **Inspection and test**: Inspections should be performed on receipt of a product and documented procedures should define the appropriate tests. Tests should be performed on all products that were repaired or serviced before they were released, to demonstrate restoration of operative condition. Records should provide evidence to demonstrate that the equipment and/or device met the necessary inspection or test criteria.
- **Inspection, measuring and test equipment**: These must be controlled, calibrated and maintained.
- **Inspection and test status**: This must be identified by using markings, stamps, labels, routing documents, inspection/test record sheets, physical location or other suitable means.
- **Control of nonconforming product**: Procedures must define the controls used to prevent the use of nonconforming product. Items should be identified, segregated and the authority for disposition made clear.
- **Corrective action**: A corrective action procedure must be documented defining what is to be analysed and how corrective actions are to be initiated and obtained to prevent re-occurrence. Corrective action procedures should also be documented for dealing with customer complaints.
- **Handling, storage, packaging and delivery**: There must be procedures for all of these. Additionally, inventories must be controlled and procedures for warranty must be written and communicated to customers.
- **Quality records**: Procedures for identification, collection, indexing, filing, storage and maintenance must be written and records must be kept.
- **Internal quality audits**: These must be planned and scheduled to verify the effectiveness of the quality system. Audits must be performed by Staff independent of the authority responsible for the area being audited. The procedures for audits, follow-up actions and reporting must be documented.
- **Training**: Procedures should be established to identify training needs.

Training must be conducted on a formal basis and records kept.

- **Servicing**: If there is a requirement to service equipment, the servicing procedures should be documented and maintained.
- **Statistical techniques**: Statistical techniques should be used where appropriate to estimate key parameters.

3.2 1981–1986

In 1981, the Department of Trade and Industry (DTI) formed a committee – FOCUS – to examine areas where standardisation could benefit the competitiveness of UK manufacturers and users of high technology – for instance Local Area Network (LAN) standardisation.

Owing to the wider international interest concerning QA, the International Organisation for Standardisation (ISO) (in conjunction with the International Electrotechnical Commission (IEC)) then set up a study group in 1983 to produce a truly *international* set of standards that all countries could use.

Reproduction of these logos courtesy of the International Organization for Standardization and the International Electrotechnical Commission

This initiative, the Open Systems Interconnection (OSI), ensured that in regard to products from different manufacturers and different countries, data could be exchanged and interworking carried out in certain defined areas. In the USA, the Corporation of Open Systems (COS) was formed in 1986 to pursue similar objectives.

3.3 1987 – ISO 9000:1987 PARTS 1, 2 AND 3

Similar to quality standards from other countries, the new ISO 9000 (1987) set of standards were very heavily based on BS 5750:1979 Parts 1, 2 and 3 [i.e. Part 1 (*Model for Quality Assurance in Design, Development, Production, Installation, and Servicing*), Part 2 (*Model for Quality Assurance in Production, Installation, and Servicing*) and Part 3 (*Model for Quality Assurance in Final Inspection and Test*)], and followed the same sectional layout except that an additional section (ISO 9000 Part 0 Section 0.1 1987) was introduced to provide further guidance about the principal concepts and applications contained in the ISO 9000 series.

When ISO 9000 was first published in 1987 it was immediately ratified by the UK, and republished by the BSI (without deviation) as the new BS 5750 (1987) standard for QMSs.

Similarly, on 10 December 1987 the Technical Board of the European Committee for Standardisation (Commission Européen de Normalisation, or CEN) approved and accepted the text of ISO 9000 (1987) as the European Standard – without modification – and republished it as EN 29000 (1987).

So, at that time official versions of EN 29000 (1987) existed in English, French and German. Other CEN members were allowed to translate any of these versions into their own language, with the same status as the original official versions.

BS 5750:1987 was, therefore, identical to ISO 9000:1987 and EN 29000:1987 except that BS 5750 had three additional guidance sections. Consequently, BS 5750 was not just the British Standard for 'Quality Management Systems', it was also *the* European and *the* international standard!

But, the question had to be, if all of these titles referred to the same quality standard, why not call the standard by the same name?

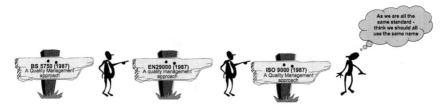

FIG. 3.8 Identical 1987 Quality Standards

3.4 1994 – BS EN ISO 9000:1994

Well, that is exactly what happened. ISO, realising the problems of calling the same document by a variety of different names was confusing (even a bit ridiculous!), reproduced (in March 1994) the ISO 9000:1994 series of documents. This series included ISO 9001 (*Model for Quality Assurance in Design, Development, Production, Installation, and Servicing*), ISO 9002 (*Model for Quality Assurance in Production, Installation, and Servicing*) and ISO 9003 (*Model for Quality Assurance in Final Inspection and Test*).

By the end of 1999, more than 60 countries had ratified ISO 9000 as their accepted Quality Management standard; Table 3.3 shows some of the national standards that were (and still are) equivalent to the ISO 9000 series.

Although the most notable change between the 1987 and the 1994 versions of the ISO 9000 standard was the streamlining of the numbering system, there were also around 250 other changes, the main ones being:

- It became an explicit requirement that all members of an organisation (down to supervisory level at least) must have job profiles (descriptions) to define their authority and responsibility.

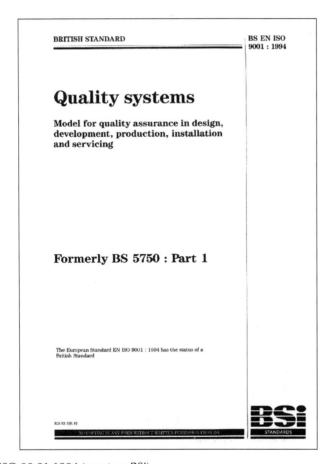

BRITISH STANDARD

BS EN ISO
9001 : 1994

Quality systems

Model for quality assurance in design, development, production, installation and servicing

Formerly BS 5750 : Part 1

The European Standard EN ISO 9001 : 1994 has the status of a British Standard

BSI
STANDARDS

FIG. 3.9 ISO 90 01:1994 (courtesy BSI)

- Design reviews became compulsory throughout the Work Package lifetime.
- Documentation control was extended to ensure that all data were up to date.

Most of these 250 changes were intended to clarify the standard, making it easier to read. However, they did not significantly alter the way in which most organisations were running their businesses, merely seeking to improve it.

3.5 2000 – ISO 9001:2000

When ISO 9000 was first released in 1987, it was recognised as being *mainly* aimed at manufacturers and it was largely incomplete, requiring auditors to fill in many gaps. The first revision of ISO 9000 in 1994 eliminated many of these problems. However, an organisation could still conform to the standard *but* at the same time produce substandard products that were of a consistently poor quality! There was clearly a major loophole that enabled organisations to comply with the

TABLE 3.3 Comparison chart of standards equivalent to ISO 9000

Standard No.	Equivalent standard								
	AS	ASQC	BS	CSA	DIN	EN	IEC	JIS	NFX
ISO 9000	AS 3900	ASQC Q90		CSA Q9000	DIN ISO 9000	EN 29000		JIS-Z9900	NFX 50–121
ISO 9000/1		AQC Q9000–1	BS EN ISO 9000–1		DIN EN ISO 9000 PT1	EN ISO 9000/1			NFX 50–121
ISO 9000/2	AS 3900.2	ASQC Q9000–2							
ISO 9000/3		ASQC Q9000–3	BS 5750 PT13 (1991)	CSA Q9000.3	DIN ISO 9000 PT3	EN 29000 PT3			NFX 50–121/3
ISO 9000/4	AS 3900.4		BS 5750 PT14 (1993)		DIN ISO 9000 PT4	EN 60300 PT1	IEC 300 PT1		
ISO 9001	AS 3901	ASQC Q9001	BS EN ISO 9001	CSA Q9001	DIN EN ISO 9001 DIN ISO 9001	EN ISO 9001		JIS-Z9901	NFX 50–131
ISO 9002	AS 3902	ASQC Q9002	BS EN ISO 9002	CSA Q9002	DIN EN ISO 9002 DIN ISO 9002	EN ISO 9002		JIS-Z9902	NFX 50–132

Equivalent standard

Standard No.	AS	ASQC	BS	CSA	DIN	EN	IEC	JIS	NFX
ISO 9003	AS 3903	ASQC Q9003	BS EN ISO 9003	CSA Q9003	DIN EN ISO 9003 DIN ISO 9003	EN ISO 9003		JIS-Z9903	NFX 50–133
ISO 9004	AS 3904	ASQC Q9004–1		CSA Q9004	DIN ISO 9004	EN 29004		JIS-Z9904	
ISO 9004/1		ASQC Q9004–1	BS EN ISO 9004–1		DIN EN ISO 9004 PT1	EN ISO 9004/1			
ISO 9004/2	AS 3904.2	ASQC Q9004–2	BS 5750 PT8 (1991)	CSA Q9004.2	DIN ISO 9004 PT2	EN 29004 PT2			NFX 50–122-2
ISO 9004/3	AS 3904.3	ASQC Q9004–3							
ISO 9004/4	AS 3904.4	ASQC Q9004–4	BS 7850 PT2 (1994)						

requirements of ISO 9001:1994 – without having to *improve* the quality of their product or service!

Some managers also found it extremely difficult to see the real benefit of having to commit more and more manpower and finance in maintaining their ISO 9000 certification, and whilst most organisations accepted that the initial certification process was worthwhile (and could result in some very real benefits), these were mainly one-offs and it was felt that once ISO 9000 had been fully adopted within an organisation, these savings could *not* be repeated.

The ISO 9000 certificate had been hanging on the wall in the reception office for many years, but there was little or no actual benefit to be gained from having to continually pay out for re-certification and surveillance fees.

On the other hand, however, BSI frequently came across organisations who initially sought ISO 9000 registration (because it was a requirement to continue business with a client), but having seen the benefits they, in turn, pushed it on down their own supply chain, thus *increasing* the requirement for ISO 9000 certification.

FIG. 3.10 The ever-increasing demand for ISO 9000 certification

So as the 1990s progressed, more and more organisations started reaping benefits from the existing ISO 9001:1994 requirements, *but* as the standard became more popular the inadequacies of ISO 9001:1994 became more apparent. For example:

- Some organisations were *not* manufacturers and did not need to carry out all 20 elements making up ISO 9000:1994 in order to be a quality organisation.
- The standard was too biased towards manufacturing industries, thus making it difficult for service industries to use.

- There was growing confusion about having three quality standards available for certification (i.e. ISO 9001, ISO 9002 and ISO 9003) and there was a need for the requirements for all three standards to be combined into one overall standard.
- The ISO 9001:1994 requirements were duplicated in many other management systems (e.g. ISO 14001 for Environmental Management and OHSAS 18001 for Occupational Health and Safety), which resulted in duplication of effort.

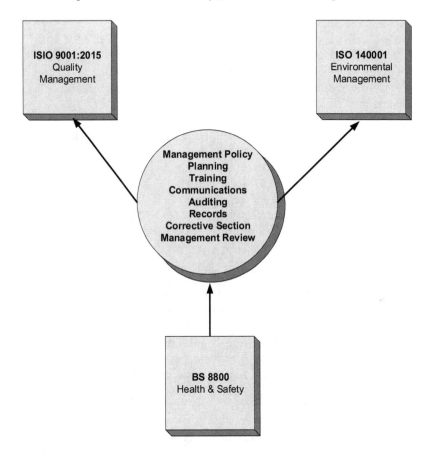

ISIO 9001:2015
Quality
Management

ISO 140001
Environmental
Management

**Management Policy
Planning
Training
Communications
Auditing
Records
Corrective Section
Management Review**

BS 8800
Health & Safety

FIG. 3.11 The common elements from quality, environmental and safety management standards

- Many organisations wanted to progress beyond the confines of ISO 9001 towards Total Quality Management (TQM).
- The documents were viewed by many as **not** being very user-friendly.
- The language used was *not* clear and could be interpreted in many different ways.
- The standard was very inflexible and could *not* be tailored to specific industries, etc.

- The standard did *not* cater for continual improvement.
- The standard did *not* fully address customer satisfaction.

The reasons went on and on and there was clearly a need for revision, with the overall aim of making a new ISO 9001:2000 that was:

- more compatible to the other management systems;
- more closely associated with business processes;
- easier to understand;
- capable of being used by all organisations, irrespective of size;
- capable of being used by all types of industry and profession (i.e. manufacturers *and* service providers);
- a means of continually improving quality; and above all
- *future proof.*

A decision was made, therefore, to provide a revised standard which would:

- be split, so that one standard (i.e. ISO 9001:2000) would address requirements whilst another (ISO 9004) would address the gradual improvement of an organisation's overall quality performance;
- be simple to use, easy to understand and use only clear language and terminology (a definite plus for most readers of current international standards!);
- have a common structure based on a 'process model';
- be capable of being 'tailored' to fit all product and service sectors and all sizes of organisation (and not just the manufacturing industry);
- be more orientated toward continual improvement and customer satisfaction;
- be capable of demonstrating continuous improvement and prevention of nonconformity;
- provide a natural stepping stone towards performance improvement;
- have an increased compatibility with other management system standards;
- provide a basis for addressing the primary needs and interests of organisations in specific sectors such as aerospace, automotive, medical devices, telecommunication and others.

ISO emphasised, however, that this revision of the ISO 9000 standards would *not* require the rewriting of the organisation's current QMS documentation! They pointed out that the only real change had been from a 'system-based' to a more 'process-orientated' management approach, which could be easily addressed by organisations on an as-required basis.

3.6 2008 – ISO 9001:2008

According to the rules of ISO (and similar to all other ISO standards), ISO 9001 is required to undergo review and revision every 5 years. The ISO 9001 standard was first revised in 1994 and then underwent a major revision in 2000. Thankfully, however, the changes made to ISO 9001:2008 were relatively minor and of little concern to most companies. The new standard did *not* contain any *new* requirements, nor did it contain changes to any of the requirements of ISO 9001:2000 and, more importantly, it did not change the objective of ISO 9001:2000.

For all intents and purposes, therefore, the structure and outline of ISO 9001:2008 was identical to that of ISO 9001:2000 and only introduced clarifications to the existing requirements based on eight years of experience of implementing the standard worldwide, with over one million certificates issued in over 175 countries. It also introduced some changes to the wording, intended to improve consistency with the other Safety (OHSAS 18001) and Environmental (ISO 14001) Management Systems.

According to ISO, the benefits of the changes to the wording in ISO 9001:2008 were as follows:

- easier to use;
- clearer language;
- easier to translate into other languages;
- better compatibility with environmental and safety management standards.

In November 2008 ISO published their latest revision, which was called (pretty obviously!) ISO 9001:2008.

3.7 2015 – ISO 9001:2015

Following the publication of the minor amendment to ISO 9001 in 2008, ISO carried out extensive research in preparation for an updated edition of ISO 9001 aimed at developing a long-term strategic plan for ISO and increasing the alignment of ISO's management system standards through the development of a common high-level structure, common definitions and some common text for *all* current and future management standards.

The main result of ISO's research indicated that whilst there was still significant satisfaction with the 2008 version of the standard, most people considered that in order to keep ISO 9001 relevant and reflect changes in its environment, a revision of ISO 9001:2008 was appropriate and that this revised standard should (among other things):

- remain generic, and relevant to all sizes and types of organisation regardless of whether they were designers, manufacturers, suppliers or end users;
- maintain the current focus on effective process management to produce cost-effective and desirable end results;
- take account of changes in QMS practices and technology since the previous major revision in 2000;
- reflect changes in the increasingly complex, demanding and dynamic environments in which organisations operate;
- enhance compatibility and alignment with other ISO management system standards;

- facilitate effective organisational implementation and effective conformity assessment by first, second and third parties;
- use simplified language and writing styles to aid understanding and consistent interpretations of its requirements; but above all
- provide a stable core set of requirements for the next 10 years or more.

Quality *used* to be about making sure that a product was right, with an emphasis on the manufacturer being required to produce something that could be inspected against a specific dimension or criterion. The product was then considered acceptable, or had to be reworked to become acceptable, or had to be scrapped (which could be very expensive). When things went wrong it was usual to blame the craftsmen (e.g. designers, software engineers, construction engineers, etc.!).

But that was in the past and we are now on the threshold of 'Integrated Management Systems' (IMS)!

FIG. 3.12
Old-fashioned Quality Control!

3.8 SO WHAT HAS CHANGED?

In a nutshell, it has:

- **brought quality and continuous improvement into the heart of every business:** The revised standard ensures that Quality Management is now completely integrated and aligned with the business strategies of your organisation.
- **emphasised the importance of leadership:** Top Management are now required to show 'hands on' leadership in motivating the whole organisation towards their organisation's goals and objectives.
- **introduced risk and opportunity management:** Reinforced use of the management system as a governance to help identify business opportunities that contribute to bottom-line improvements.
- **introduced a new integrated approach to management standard:** With the new structure detailed in Annex SL (previously known as ISO Guide 83) forming the basis for all generic management systems, it has now become mandatory for *all* new and newly revised ISO management systems standards to abide by this structure, so that in future it will be much easier to implement multiple, integrated management systems.

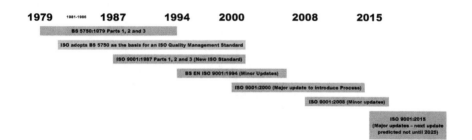

FIG. 3.13 The history of the ISO 9001 Standard

Author's Note

As you can see from this brief historical overview, the management of quality has increased from small beginnings way back in the 13th century to something that now effects virtually all organisations whether they are designers, manufacturers, suppliers, installers or end users. But the next question is 'who produces all of these rules and regulations that now apply'?

In Chapter 4 we shall have a look at the various national and international organisations producing these standards that we have come to accept as an everyday way of life.

Who produces quality standards?

CONTENTS

Author's Note

Now that we have seen how quality standards have developed over the last eight centuries or so, the next question is who actually produces these standards?

In Chapter 4 you will see that there are many Standards Bodies operating in the world today and although the International Standards Organisation (ISO) is by far and away the largest one producing industrial standard, there are two other very important organisations who have been heavily involved in the production of standards for electronic and electrical related technologies – i.e. the International Electrotechnical Commission (IEC) and the International Telegraph Union (ITU) for information and communication standardisation – which we shall now have a look at.

There are three global sister organisations (ISO, IEC and ITU) that develop international standards for the world and, when appropriate, joint committees work together to ensure that international standards fit together seamlessly and complement each other.

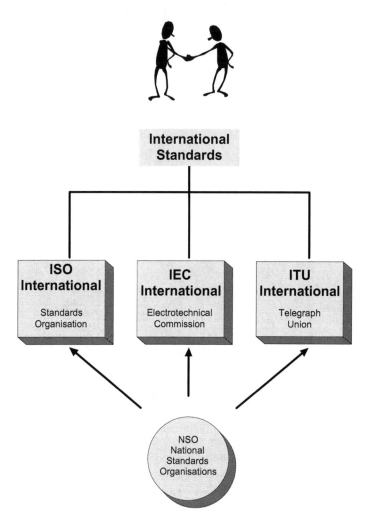

FIG. 4.1 International standards

The international standards are, themselves, drawn up by a series of inter-national technical committees which have been approved by ISO or IEC member countries, and there are now many hundreds of different ISO and IEC standards available, covering virtually every situation.

Due to the steady growth in international standardisation, ISO and the IEC are now the standards bodies to which most countries are affiliated – via, that is, their own particular National Standards Organisation (NSO).

Although these standards were initially published as 'recommendations', they are now globally accepted as international standards in their own right and the use of the words 'must' and 'shall' (i.e. denoting a mandatory requirement) has become commonplace.

4.1 WHO AND WHAT IS ISO?

FIG. 4.2 ISO founder members, London 1946

Photo courtesy of the International Standards Organisation

In 1946, 25 delegates from 25 different countries met at the Institute of Civil Engineers in London and decided to create a new international organisation 'to facilitate the international coordination and unification of industrial standards'.

As a result, the International Standards Organization (ISO) was established as a United Nations Agency in 1947 and is an independent, non-governmental membership organisation which (at the time of writing) is made up of member bodies from 162 countries with one standards body representing each member country (e.g. BSI for the United Kingdom and ANSI for the United States). These representatives make up a number of Technical Bodies (such as ISO/TC 176 for Quality Management and Quality Assurance) that are responsible for the development of standards. ISO has a Central Secretariat based in Geneva, Switzerland and is the world's largest producer of voluntary international standards covering almost all aspects of industry and manufacturing.

FIG. 4.3 The International Standards Organization logo

Reproduction of logo courtesy of the International Organization for Standardization

---

☼ Author's Note

Because the 'International Organization for Standardization' would have different acronyms in different languages [IOS in English, OIN in French (*Organisation internationale de normalisation*)], the founders decided to give it the short form 'ISO' which is derived from the Greek *isos*, meaning equal. Thus, whatever the country, whatever the language, ISO standards are always equally applicable and equally acceptable.

There are three member categories of ISO. Each enjoys a different level of access and influence over the ISO system, which helps ISO to be inclusive while also recognising the different needs and capacity of each national standards body. There are (at the time of writing):

- **162 Full members** (or member bodies) who influence ISO standards development and strategy by participating and voting in ISO technical and policy Meetings. Full members sell and adopt ISO international standards nationally.
- **38 Correspondent members** who observe the development of ISO standards and strategy by attending ISO technical and policy Meetings as observers. Correspondent members can also sell and adopt ISO International Standards nationally.
- **5 Subscriber members** who keep up to date on ISO's work but cannot participate in it. They do not sell or adopt ISO international standards nationally.

Since 1947, ISO has published more than 19,500 international standards covering almost every aspect of technology and business. From food safety to computers, and agriculture to healthcare, it is certainly true to say that ISO international standards impact all of our lives!

4.2 WHO AND WHAT IS THE IEC?

FIG. 4.4 The International Electrotechnical Commission logo

Reproduction of logo courtesy of the International Electrotechnical Commission logo

Founded in 1906, the IEC (International Electrotechnical Commission) is a not-for-profit, non-governmental organisation responsible for the preparation and publication of international standards for all electrical, electronic and related technologies – collectively known as 'electrotechnology'.

FIG. 4.5 First council meeting of the IEC in London

Photo courtesy of the International Electrotechnical Commission

Currently, 60 countries are Full members while another 23 participate in the Affiliate Country Programme (AFS) which, although not really a form of membership, is nevertheless designed to help industrialising countries get involved with the IEC.

IEC enables companies, industries and governments to meet, discuss and develop international standards that are relevant to millions of devices that contain electronics and use (or produce) electricity – and which in turn rely on IEC international standards and conformity assessment systems to perform, fit and work safely together.

The IEC:

- provides a standardised approach to testing and certification;
- helps make testing transparent, predictable, comparable from one country to the next and more affordable;
- provides a single set of rules and procedures which prevents unfair advantages for local products in a particular country;
- allows companies, governments and test labs to verify that a certain product or system conforms to the requirements that are described in a particular standard.

4.3 WHO AND WHAT IS THE ITU?

FIG. 4.6 The International Telecommunication logo

Reproduction of logo courtesy of the International Telecommunication Union

Originally founded in 1865 as the International Telegraph Union, the ITU is the oldest existing international organisation, with its headquarters in Geneva, Switzerland. The ITU [as a specialised agency of the United Nations (UN)] is responsible for *all* issues that concern information and communication technologies, and it helps undeveloped countries to establish and create telecommunication systems of their own.

The ITU sets and publishes regulations and standards relevant to electronic communication and broadcasting technologies including radio, television, satellite, telephone and the Internet. Although the recommendations of the ITU are non-binding, most countries adhere to them in the interest of maintaining an effective international electronic communication environment.

As well as these ITU regulations, their committees [the Radiocommunication Sector (ITU-R) and the Telecommunication Development Sector (ITU-D)] also publish recommendations.

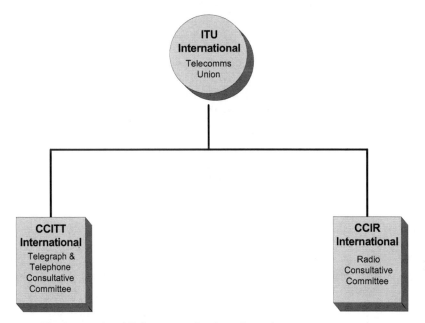

FIG. 4.7 The International Telecommunications Committee

4.4 HOW IMPORTANT IS THE WORK OF ALL THESE COMMITTEES?

From the consumer's point of view, the importance of international (i.e. ISO and IEC) standardisation is that all major agencies are now committed to recognising these standards. Equipment, modules and components can thus be designed and built so that they will be acceptable to all member countries, and today there is a constant demand for new, revised and updated standards – particularly those with an international relevance. A good example is, of course, ISO 9001, which (as the most generically successful and widely used QMS ever devised) has now become the benchmark for improving business efficiency and competitiveness worldwide.

It must not be forgotten, however, that the overall aim of standardisation is not just to produce paperwork that becomes part of a library. The aim is to produce a precise, succinct, readily applied and widely recognised set of principles that are relevant and satisfy the varied needs of business, industry or commerce.

The aim is also that standardisation shouldn't provide exclusive advantage to the products or services of one particular individual supplier and that the application of standards should always be capable of being verified by an independent third party evaluator (i.e. an auditor).

4.5 MILITARY STANDARDS

In the military world the situation is little different. The United Kingdom Ministry of Defence (MOD-UK) use Defence Standards (DEF STANS), the American Division of Defense (DOD) use Military Standards (Mil-Std), the North Atlantic Treaty Organization (NATO) use NATO Allied Quality Assurance Publications (AQAPs) and most other nations have their own particular variations.

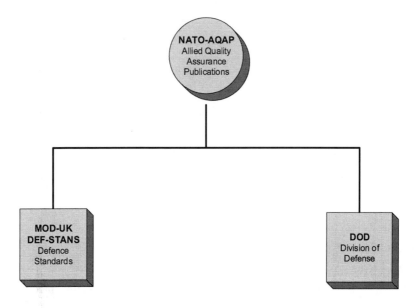

FIG. 4.8 Military standards

4.6 NATIONAL STANDARDS

A standard is a document that provides requirements, specifications, guidelines or characteristics that can be used consistently to ensure that materials, products, processes and services are fit for their purpose. With literally thousands of standards available worldwide, on every conceivable topic, this can make it difficult to decide which is important to you!

It has to be said, however, that standards are as international as the markets they serve and currently the main producers of national standards in Western Europe are:

- United Kingdom: British Standards Institution (BSI);
- Germany: Deutsch Institut fur Normung e.v. (DIN);
- France: Association Français de Normalisation (AFNOR).

QUALITY IS WORLDWIDE

FIG. 4.9 National standards

Courtesy Stingray Quality Management Consultants

Outside Europe the most widely used standards are:

- USA: American National Standards Institute (ANSI);
- Canada: Canadian Standards Association (CSA).

There are others, of course (e.g. Japan and Saudi Arabia), but Europe and North America are the main continents where such standards are used.

Nevertheless, although these countries publish what are probably the most important series of standards, virtually every country with an industrial base has its own national organisation producing its own national set of standards. However, national bodies and national standards cannot dictate customer choice. A product that may legally be marketed need not be of universal appeal or internationally acceptable (e.g. the three-pin electrical plug used in the UK is totally useless in

FIG. 4.10 Main producers of national standards outside Europe

most other countries!) and this diversity of standards can obviously lead to a lot of confusion, especially with regard to international trade and tenders.

For example, if the USA were to invite tenders for a project quoting ANSI national standards as the minimum criteria, a European organisation could find it difficult to submit a proposal, either because it didn't have a copy of the relevant standard or they wouldn't find it cost effective to retool their entire works in order to conform to the requirements of that particular US domestic standard.

Indeed, where different national standards persist, they will do so as a reflection of different market preferences and national idiosyncrasies. For industry to survive in this new, 'liberalised' market, therefore, it must have a sound technological base supported by a comprehensive set of internationally approved standards.

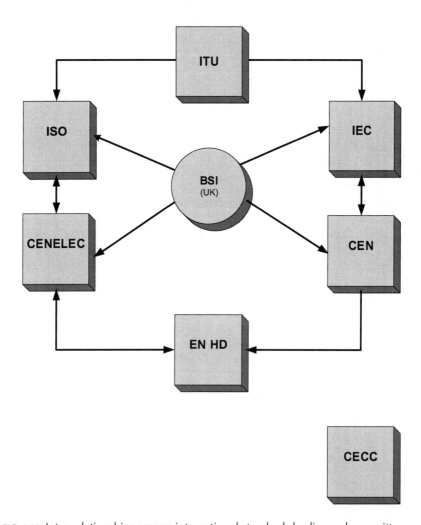

FIG. 4.11 Inter-relationships among international standards bodies and committees

'Quality' has thus become the key word in today's competitive markets

In the UK, BSI is responsible for the actual production of national standards and its committees correspond closely with those of other European and/or international standards organisations. BSI also presents the UK viewpoint to the European standards organisations CEN and CENELEC (European Committee for Electrotechnical Standardisation), as well as ETSI (European Telecommunications Standards Institute) in the telecommunications field. These organisations seek to develop harmonised European standards that are crucial to the success of the European Single Market. In the broader international arena, ISO and IEC pursue similar aims towards harmonising world standards. Again, BSI is active in ensuring that the views of UK businesses are represented.

 It should be remembered, however, that a proposal for a new standard can be made by any country, or indeed anybody, and will be investigated depending on the support it can attract and, critically, its ability to provide an initial draft within a workable deadline.

In all, there are more than 8000 IEC standards and guides, and since the 1970s BSI has published most of these as British standards with a national foreword. Under European agreement, BSI also publishes European standards (EN numbers) as identical British standards (e.g. EN 14988–1:2006, regarding the safety of children's high chairs, is published in the UK by BSI as BS EN 14988–1:2006, with a national foreword).

Author's Note

Hopefully these last four chapters have given you a reasonable overview of what Quality Management is all about and how the requirement for a Quality Management System has now become a worldwide normal way of life – so now let's have a look at the 'beast' we call ISO 9001:2015.

A brief summary of ISO 9001:2015

CONTENTS

> ### Author's Note
>
> Now (assuming that you're still happily reading through this book) it is time to get down to the nitty gritty of ISO 9001 and to learn a bit more about the background to this standard, how it has grown and developed over the years and how; with the introduction of 'Annex SL', it has a similar structure to other management system standards.

As previously touched on, ISO 9001:2015 is an International standard of requirements against which a company's Quality Management System can be evaluated. It is adaptable and non-prescriptive, allowing it to be applied to any type of business. It is the most widely recognised standard in the world and over a million organisations are certified to it globally, with many more certifying to sector-specific versions of the standard. It has become an international by-word for quality, with organisations reporting that implementation leads to improved efficiency, increased profits and, most importantly, greater customer satisfaction.

5.1 BACKGROUND REMINDER

When ISO 9000 was first released in 1987, it was recognised as being largely incomplete and the auditors were required to fill in lots of the gaps. The first revision of the ISO 9000 series, in 1994, was a definite step forward in the right direction but an organisation could still conform to the standard while at the same time produce substandard products and services – but at least they were of a *consistently poor* quality!

FIG. 5.1

The background to ISO 9001:2015

Even though the 1994 edition addressed most of the previous problems with the 1987 standard, by introducing 20 specific – but separate – requirements (e.g. document control, review procedures, customer satisfaction, etc.) for an organisation to prove compliance, meant that any internal or external audit was virtually a 'tick in the box' exercise.

Even more contentious was the problem that some organisations were not manufacturers and, therefore, did not need to carry out all the 20 elements making up the ISO 9001:1994 standard in order to become a quality certified organisation. It also gave organisations a choice of three 1994 standards (i.e. ISO 9001, ISO 9002 and ISO 9003) to choose from – and the feeling in some corners of the industrial world was 'why bother with it all?!'

There was clearly a need for revision, with the overall aim of making a new ISO 9001 standard that was:

- more compatible with the other management systems;
- more closely associated with business processes;
- easier to understand;
- capable of being used by all organisations, no matter their size;
- capable of being used by all types of industry and profession (i.e. manufacturers **and** service providers).

These aims were included in the 2000 version of ISO 9001 (which incidentally did away with the three-sector format of the previous 1994 standard and became a combined ISO 9001: 2000), and this new standard was immediately welcomed by all sectors of business as a step in the right direction – particularly as it emphasised the need for a business process format.

Following the publication of some minor amendments to ISO 9001 in 2008 and, in accordance with the rules of ISO, all international standards need to be reviewed and revised every 5 years and so in 2015 a new version of ISO 9001 standards was published but this time, in accordance with the rules set out in Annex SL (see Chapter 5.5) and in a similar format to all future ISO management standards.

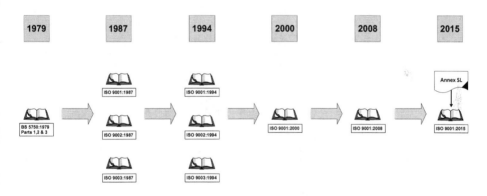

FIG. 5.2 The history of ISO 9000 so far

5.2 WHY HAS ISO 9001 BEEN REVISED?

Similar to other ISO standards, ISO 9001 has to be reviewed every five years to make sure that it is still current and relevant in the marketplace. During this revision process, the 2015 edition of ISO 9001 took into consideration the new rules regarding the common structure (with standardised core text, structure and definitions) that all new and revised management systems must in future be written to. The aim of the new ISO 9001:2015 standard, therefore, was not only to integrate with these other management system standards but also to:

- maintain ISO 9001's relevance;
- provide a consistent foundation for the long term;
- increase the adoption of ISO 9001;
- focus on an increased variety of business users (e.g. service industries, office environments);
- address the increasing complexity of the business environment;
- provide an effective solution for the non-office/virtual office environment;
- focus on achieving value for the organisation and its customers; and in doing so
- enhance an organisation's ability to satisfy customers.

Author's Hint

Although the text and layout has changed, it should be noted that the actual requirements of ISO 9001 mainly remain the same and are much less prescriptive.

5.3 HOW ARE THESE STANDARDS REVISED?

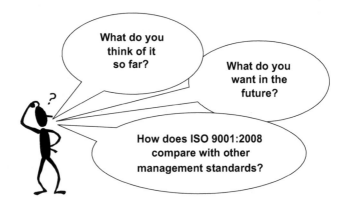

FIG. 5.3 The ISO/TC 176 survey of ISO 9001:2015

The revision process was the responsibility of ISO/TC176 (ISO Technical Committee No 176 'Quality Management and Quality Assurance'), which followed the normal work-flow used by ISO as shown below:

Initial specifications and goals were established following extensive user surveys to ensure that the standards produced would actually meet the requirements of the user, and these were followed by a user verification and validation process.

Author's Hint

Once a draft international standard has been adopted by the technical committee, it is then circulated to member bodies for voting. Publication as an international standard then requires a two-thirds majority of the votes.

FIG. 5.4 The revision process of an international standard

5.4 WHAT EFFECT DID THIS REVISION HAVE ON ISO 9001?

As stated by ISO Technical Committee No 176:

- The revised standard provides a stable, core set of requirements for the next 10 years or more.
- The standard's requirements remain generic, but are now relevant to organisations of all types and sizes operating in any sector.
- While maintaining its current focus on effective process management, the standard also reflects the changes in Quality Management System practices and technology since the last major revision in 2000.
- By applying Annex SL, the revised standard improves compatibility and alignment with other ISO management system standards.
- ISO 9001:2015 improves implementation and conformity assessments for first, second and third parties.
- Using simplified language and writing styles, the new updated standard will help all stakeholders to understand and interpret key areas better.

Author's Hint

These changes are intended to clarify the intent of a particular clause, or make implicit requirements more explicit.

Whilst this should not require an organisation to dramatically change its existing management system in order to retain their ISO 9001 certification in the future (or for a small business to continue to work in compliance with the standard), it is recommended that organisations should nevertheless take this opportunity to review whether the new wording of the standard is supported by their existing organisational practices.

5.4.1 Other changes

- **Structure and terminology**: the clause sequence and some of the terminology have been changed to improve alignment with other management systems standards as shown in Table 5.1.

Author's Hint

There is *no* mandatory requirement for the terms used by any organisation in *their existing* Quality Management System to be replaced by the terms used in this international standard! Organisations can choose to use terms that suit their own particular operations (e.g. they can use 'records', 'documentation' or 'protocols' rather than 'documented information'; and 'supplier', 'partner' or 'vendor' rather than 'external provider').

Organisations may also have someone called the Quality Manager to oversee their Quality Management System (which can be contained in an overarching document called a 'Quality Manual') if they so wish.

So don't worry, you won't get a nasty letter from ISO if you do!!

TABLE 5.1 ISO 9001:2008 vs. ISO 9001:2015

ISO 9001:2008	ISO 9001:2015
Products	Changed to 'Products and services'
Exclusions	No longer used
Management representative	No longer used
Documentation, Quality Manual, documented procedures, records	Now referred to as 'Documented information'
Work environment	Changed to 'Environment for the operation of processes'
Monitoring and measuring equipment	Changed to 'Monitoring and measuring resources'
Purchased product	Changed to 'Externally provided products and services'
Supplier	Changed to 'External provider'

- **Products and services**: Previously ISO 9001:2008 used the term 'product' to include all output categories. This has always been a bone of contention with many non-manufacturing companies and so ISO 9001:2015 has now changed this term to 'products and services' to include all output categories (hardware, services, software and processed materials).
- **Understanding the needs and expectations of interested parties**: Organisations need to demonstrate their ability to consistently provide products and services that meet customer and applicable statutory and regulatory requirements.
- **Risk-based thinking**: Previously ISO 9001:2008 had a separate clause for 'preventative action'. ISO 9001:2015 now requires an organisation to determine risk as a basis for planning and implementing Quality Management System processes and overall improvements.
- **Applicability**: Although ISO 9001:2015 no longer contains 'exclusions' regarding the applicability of its requirements to the organisation's Quality Management System, organisations can still review the applicability of the standard's requirements with regard to the size/complexity of their organisation, the management model it adopts, the range of the organisation's activities and the nature of the risks and opportunities it encounters.
- **Documented information**: In alignment with other management system standards, the term 'documented information' is now used in ISO 9001:2015 for *all* document requirements (e.g. 'document', 'documented procedures', 'Quality Manual', 'Quality Plan' or 'records').

Author's Hint

As previously mentioned, this does not, of course, prevent organisations from continuing to use these terms if they so wish, however!

- **Organisational knowledge**: To ensure the operation of its processes and so that it can achieve conformity of products and services, ISO 9001:2015 emphasises the need for organisations to determine, manage and maintain their knowledge base.
- **Control of externally provided processes, products and services**: It is now recommended that organisations apply risk-based thinking in order to determine the type and extent of controls appropriate to particular external providers and externally provided processes, products and services.

5.5 WHAT IS ANNEX SL?

But what exactly is this Annex SL thing that everyone is talking about? In short, it's a rule book for standard writers setting out what they must and must not do in the development of any international standard.

As well as ISO 9001 for QMSs, ISO also publishes (see Chapter 6) a number of other Management Systems Standards (MSS) (e.g. Environmental, Occupational Health & Safety, Risk, Business Continuity, Information Security, Asset Management Systems) and, of course, there are many other standards that are actually *based* on ISO 9001 (such as ISO/IEC 27001 for information security management systems,

FIG. 5.5
Is Annex SL going to be of any use to me?

etc.). Nevertheless, the structure and format of each of these standards is currently very different and so it is not surprising, therefore, that these variations continue to cause problems for organisations who have to become certified against multiple schemes.

☀ Author's Hint

A good example of this was a company who were manufacturing therapeutic medical devices which, as they concerned patient safety, would have to be assessed against both ISO 9001 (for Quality Management) and ISO 13485 (for medical devices). The company asked for my assistance back in the late 1990s and this was the first time that I realised how very similar were the requirements between the two standards, and so I was able to assist them in developing a Quality Management System that covered the requirements of both standards and which could be audited by a notified body, thereby reducing both time and costs for them.

Thankfully ISO has now also realised the potential problems and, in an attempt to produce some form of consistency across the standards, it has agreed that in future all new revisions to existing MSS must have the same high-level structure [as detailed in Appendix 3 of ISO/IEC Directives, Part 1 Annex SL (commonly referred to as just 'Annex SL')] identical core text, as well as common terms and definitions where suitable (see example below) – with the overall aim of making these easier to read and understand and, thus, easier to integrate more than one standard into an overall business management system.

Whilst the high-level structure cannot be changed, sub-clauses and discipline-specific text can be added. This commonality will, in turn, require less maintenance and audit resources, meaning that it can be changed more easily to meet evolving business needs.

Top management shall ensure that the responsibilities and authorities for relevant roles are assigned and communicated within the organization.

FIG. 5.6 Example of identical text across different standards

In adopting Annex SL, ISO considered that new standards could be designed to be much less 'fussy', allowing organisations greater freedom to design their own management system and able to create organisation-specific manuals and procedures.

As we all know, external audits can become very disruptive but the increased commonality of requirements across standards will mean that there is potential for a change in the way that third party assessments can be completed. For example, if an organisation normally has a Health and Safety audit in January, a Quality Management audit in May and an Environmental Management audit in September, they could now combine all these into one single audit – thus saving time and money!

By setting up this identical high-level structure using identical clause titles (and sequence of clause titles), text definitions, scope, common references, terms and definitions, organisational structure, leadership and planning support, ISO has ensured that at least 30% commonality of text exists between each management system.

FIG. 5.7 Conflicting external audits

Annex SL applies to all management system standards [such as full ISO standards, Publicly Available Specifications (PAS) and Technical Specifications (TS)], and two of ISO's flagship management system were published in 2015 (ISO 9001 and ISO 14001), with both using the new format for their revisions. In addition, a number of other standards [such as ISO 27001 (Information security management), ISO 22301 (Business continuity management) and ISO 22301 for Business continuity management)] already employ Annex SL.

So, with the introduction of Annex SL, all of the major clause numbers and titles of *all* ISO management system standards will now be identical, such as the introduction, scope and normative references (whose content will be specific to each discipline), terms and definitions (22 terms and definitions have been listed which must be addressed; they cannot be deleted or changed – however, each standard can *add* their own additional terms and definitions if required and also add to or modify the notes written against these stated terms and definitions) and operation. ISO has, however, said that as an option, each standard may have its own bibliography!

5.5.1 What is the new structure of Annex SL?

The structure of Annex SL now concerns 10 main clauses instead of the previous 8 used by ISO 9001:2008 and other similar management standards, as shown below:

- **Clause 1: Scope:** The scope sets out the intended results of the management standard (see Clause 4).

- **Clause 2: Normative references:** Provides details of the reference standards or publications relevant to the management standard.

- **Clause 3: Terms & definitions:** Details terms and definitions applicable to the management standard in addition to any formal related terms and definitions.

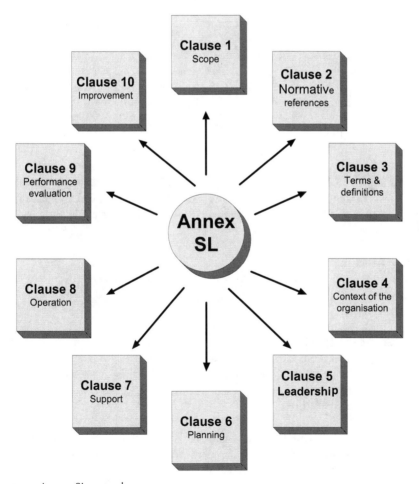

FIG. 5.8 Annex SL core clauses

- **Clause 4: Context of the organisation:** Clause 4 is the basic foundation of any management system. It will determine:
  - why the organisation is here;
  - how the organisation is structured;
  - which internal and external issues can impact on its intended outcomes;
  - who are their potential clients and what are their requirements. This in turn will need to be documented in the scope of the particular standard, which will then set the boundaries of the management system – i.e. 'what's in' and 'what's out'.

- **Clause 5: Leadership:** The new high-level structure places particular emphasis on leadership, not just management as set out in previous standards. This means that Top Management will now:
  - have greater accountability and involvement in the organisation's management system;
  - need to integrate the requirements of the management system into the organisation's Core Business Process;
  - need to ensure the management system achieves its intended outcomes;
  - need to allocate the necessary resources. Top Management is also responsible for communicating the importance of the management system and heightening employee awareness and involvement.

- **Clause 6: Planning**

 Risk-based thinking is probably *the* most important requirement for the new version of ISO 9001, and once an organisation has highlighted the risks and opportunities in their product or service, they will then need to decide how and when these risks will have to be addressed and, most importantly, by whom.

Author's Hint
This new risk approach is a major change to management systems and replaces preventative action which will, in turn, reduce the need for corrective actions later on.

- **Clause 7: Support:** Having thought about their overall organisational aims and policies, commitment and planning, organisations will now have to consider what resources are required to meet their goals and objectives.

- **Clause 8: Operation**

Clause 8 contains the lion's share of management system requirements, addressing:
  - in-house and outsourced processes;
  - overall process management;

- – sufficient criteria to control these processes; and
- – ways to manage planned and unintended change.

- **Clause 9: Performance evaluation**: Clause 9 determines:
  - – what, how and when things are to be monitored, measured, analysed and evaluated;
  - – how, when and by whom internal audits are conducted so as to ensure that the management system conforms to the requirements of the organisation as well as the selected management standard;
  - – how the organisation's management system is successfully implemented and maintained;
  - – how management reviews are organised and conducted to determine whether they are, and can remain, suitable, adequate and effective.

- **Clause 10: Improvement**: It is a well-known fact that in an ever-changing business world, not everything will always go according to plan and so Clause 10 looks at ways to address nonconformities and corrective action, as well as suggesting strategies for improvement on a continual basis.

Author's Hint

ISO management system requirements were, of course, not written for the express purpose of audit and certification! However, the development of an agreed format will definitely make internal, external and third party auditing far easier for both the auditors and the organisation.

5.5.2 What is the current status of Annex SL revisions?

Although ISO emphasises that Annex SL has made its associated manage- ment standards less a series of 'requirement standards' (except for standards like ISO 13485, which has to have some mandatory requirements because of patent safety), Annex SL, nevertheless, by introducing identical core text, has in turn led to 45 'shall' statements – generating 84 requirements which have had a knock-on effect to ISO 9001:2015, which now has *over 130 'shall' requirements!*

The impact of this revision has been similar to, if not greater than, the 2000 edition, which was a major change for accreditation bodies, Certification Bodies, training organisations, implementing organisations, procurement organisations, consultants and customers.

Organisations may now have to align their management systems with the structure of the revised standard, for example:

- The organisation's Quality Manual may no longer be required but, at the very least, the associated processes, procedures and Work Instructions, etc. will probably need to be amended and retained as documented information.
- A risk management processes may need to be developed to determine the level and extent of control for the 'external provision of products and services' if not already in place.

This will have implications for the organisation's procurement and outsourcing activities and, therefore, has implications for suppliers.

• Internal auditors will need to become familiar with the revised ISO 9001:2015 standard and so training may need to be considered.

At first glance, Annex SL appears to make the standard writers' lives 'much easier' but in reality, as organisations begin to understand and appreciate the value of different management systems all speaking a common language, it will be organisations – and in their turn, the consumer – who stand to be the true beneficiaries.

So, with the introduction of this new Annex SL, the various ISO technical committees will have a lot to do! But having said that, with a harmonised structure, text, terms and definitions, this leaves the standard's developers with a certain amount of flexibility to incorporate their specific technical topics and requirements.

5.6 BUT WHAT EXACTLY IS MEANT BY THE 'ISO 9000 FAMILY OF STANDARDS'?

FIG. 5.9 The ISO 9000 family

The ISO 9000 family of standards consists of three primary standards supported by a number of technical reports. These are:

- ISO 9000:2015 Quality Management Systems: fundamentals and vocabulary, which describes the fundamentals of QMSs and specifies the terminology for QMSs.
- ISO 9004:2009 Managing for the sustained success of an organisation: a Quality Management approach which provides guidance on QMSs, including the processes for continual improvement that will contribute to the satisfaction of an organisation's customers and other interested parties.
- ISO 9001:2015 Quality Management Systems: requirements, which is the most important Quality Management *requirements* standard that is applicable to all organisations, products and services.

5.6.1 ISO 9000:2015 Quality Management Systems: fundamentals and vocabulary

FIG. 5.10 The way to ISO 9000:2015

To ensure a more harmonised approach to standardisation (and hopefully, achievement of coherent terminology within the ISO 9000:2015 family!), ISO 9000:2015 was developed to assist:

- those concerned with enhancing the mutual understanding of the terminology used in quality (e.g. producers, suppliers, customers, regulators);
- internal or external auditors, regulators, certification and/or registration bodies;
- developers of related standards;
- organisations that provide advice or training on quality matters.

ISO 9000:2015 also provides:

- an introduction to the fundamentals of Quality Management Systems;
- terms and definitions;
- the methodology used in the development of the vocabulary.

5.6.2 ISO 9004:2009, managing for the sustained success of an organisation: a quality management approach

FIG. 5.11 The reason for ISO 9004:2009

This is the fourth edition of the standard which now contains up-to-date information on achieving sustainable business success through Quality Management. It provides guidance on QMSs, including the processes that are required for continuous improvement and, ultimately, customer satisfaction. It outlines the importance of self-assessment to identify areas of strength within organisations and areas where improvements can be made.

Author's Hint

ISO 9004, however, is not intended to provide certification or regulatory requirements, but it will help you achieve and maintain business objectives over the long term. It builds on ISO 9001 to widen the scope of your quality management and give you greater confidence in how you assess, maintain and improve it.

ISO 9004 outlines a systematic approach for continual improvement of your organisation's overall performance. This includes best practice advice on quality strategy and policy, as well as managing resources and processes.

ISO 9004:2009 also provides advice regarding:

- Quality Management principles;
- managing for the sustained success of an organisation;
- strategy and policy;
- resource management;
- process management;
- monitoring, measurement, analysis and reviews;
- improvement, innovation and learning;
- self-assessment tools;
- normative references;
- terms and definitions.

5.6.3 ISO 9001:2015 Quality Management Systems: requirements

FIG. 5.12 The benefits of gaining ISO 9001:2015

ISO 9001:2015 is a single Quality Management *'requirements'* standard that is applicable to all organisations producing products and/or services. It is the only standard that can be used for the certification of a QMS, and its generic requirements can be used by *any* organisation to:

- address customer satisfaction;
- meet customer and applicable regulatory requirements;
- enable internal and external parties (including Certification Bodies) to assess the organisation's ability to meet these customer and regulatory requirements.

For certification purposes, your organisation will now have to possess a documented management system which takes the inputs and transforms them into targeted outputs – something that effectively:

- says what you are going to do;
- does what you have said you are going to do;
- keeps records of everything that you do – especially when things go wrong.

The basic process to achieve these targeted outputs will encompass:

- client requirements;
- inputs from management and Staff;
- documented controls for any activities that are needed to produce the finished article; and, of course
- delivering products and services which satisfy the customer's original requirements.

5.7 THE SEVEN PRINCIPLES OF MANAGEMENT

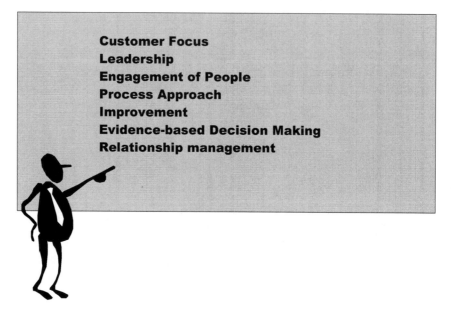

Customer Focus
Leadership
Engagement of People
Process Approach
Improvement
Evidence-based Decision Making
Relationship management

FIG. 5.13 The seven Quality Management principles

As stated in the standard, ISO 9001:2015 is the internationally recognised standard for *all* Quality Management Systems. As such, it provides the benchmark against which a company's management system is measured and, if found to be adequate, certified as compliant. Consequently, for this standard to be seen as the leading example of management system methodology, it has had to reflect currently accepted best practice.

With this in mind, the standard was developed to be totally business focused, and aimed at improving an organisation's management system through the application of seven proven principles, as shown below.

☀ Author's Hint

ISO 9001:2008 was previously based on eight Quality Management Principles.

With the introduction of Annex SL (and the omission of the 'A systems approach to management'), ISO 9001:2015 is now based around seven Quality Management principles, and the rather confusing phrase 'Mutually beneficial supplier relationships' has thankfully now been replaced by a far simpler phrase, 'Relationship Management'.

The seven Quality Management principles:

1 Customer focus

Seeking to satisfy the demands and
expectations of the customer. Organisations
depend on their customers and should,
therefore:

- understand current and future customer
  needs;
- meet customer requirements (and their
  budget!);
- strive to exceed customer expectations.

*The primary focus of Quality Management is to
meet customer requirements and to strive to
exceed customer expectations.*

Annex SL

2 Leadership

- Leadership is a core skill in today's
  business world; without strong leadership,
  many good businesses fail – and so can
  management systems.
- Leaders create the environment in which
  people can become fully involved in
  achieving the organisation's objectives.
- Without the right support and direction
  for your organisation you can't achieve
  your full potential.

*Leaders at all levels establish unity of purpose
and direction and create conditions in which
people are engaged in achieving the quality
objectives of the organisation.*

Annex SL

3 Engagement of people

- 'Employee engagement' should aim at
  ensuring that all employees are committed
  to their organisation's goals and values.
- People at all levels are the essence of an
  organisation, and their full involvement
  enables their abilities to be used for the
  organisation's benefit.

It is essential for the organisation that all people are competent, empowered and engaged in delivering value.

Annex SL

4 Process approach

- The logical sequencing of activities to efficiently achieve a desired result.
- A desired result is achieved more efficiently when related resources and activities are managed as a process.

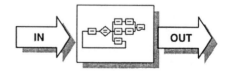

Consistent and predictable results are achieved more effectively and efficiently when activities are understood and managed as interrelated processes that function as a coherent system.

Annex SL

5 Improvement

- It is essential for an organisation to maintain current levels of performance, to react to changes in its internal and external conditions in order to create new opportunities.
- The improvement of any organisation's Quality Management System and processes, etc. should become a permanent objective. Put simply, it means 'getting better all the time'.

Successful organisations have an ongoing focus on improvement.

Annex SL

6 Evidence-based decision making

- Decision making based on the analysis of data should always be based on facts and evidence.
- Effective decisions are based on the logical and intuitive analysis of data and information.
- Reliable data gathered via planned measures is the only way this can be achieved.

So where are our problems?

Decisions based on the analysis and evaluation
of data and information are more likely to
produce desired results.

Annex SL

7 Relationship management

- The mutual support of an organisation and its suppliers adds value.
- Mutually beneficial relationships between an organisation and its suppliers enhance the ability of both organisations to create value.
- The supplier/customer relationship should always be viewed as an *interdependent* relationship.

For sustained success, organisations manage their relationships with interested
parties, such as suppliers.

Annex SL

5.7.1 The Process Model

The seven management principles are reflected in the ISO 9001 process model shown below, indicating how each of these principles is embraced within the ethos of the standard.

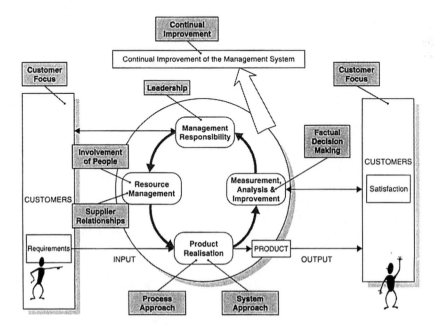

FIG. 5.14 The ISO 9001:2015 process model and the seven management principles

5.8 WHAT IS THE NEW STRUCTURE OF ISO 9001?

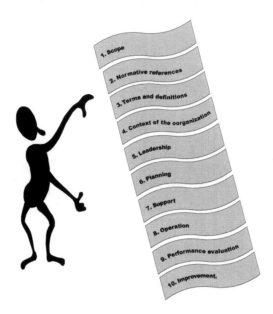

FIG. 5.15 The major clause numbers

In compliance with Annex SL, the structure and format of the ISO 9001:2015 standard has changed from the previous 8 main clauses to 10, as follows:

Clause 1: Scope

Demonstration that the organisation has the ability to supply products and services that consistently meet both customer and applicable statutory and regulatory requirements.

Clause 2: Normative references

For ISO 9001:2015, there are no references other than those stated in Annex SL for all management system standards.

Clause 3: Terms and definitions

A description of the standard-specific terms and definitions.

Clause 4: Context

- Understand your organisation and its unique context.
- Clarify the needs and expectations of interested parties.
- Define the scope of your QMS.
- Establish a QMS that complies with this standard.

Clause 5: Leadership*

Provide leadership by:

- focusing on quality and customers;
- establishing a suitable Quality Policy;
- defining roles and responsibilities.

Clause 6: Planning

Risks and opportunities:

- Consider risks and opportunities when you plan your QMS.
- Plan how you're going to manage risks and opportunities.
- Define actions to manage risks and address opportunities.

Quality objectives:

- Set quality objectives and develop plans to achieve them.
- Establish quality objectives for all relevant areas.
- Develop plans to achieve objectives and evaluate results.
- Control changes to your QMS.

Clause 7: Support

Support your Quality Management System by:

- providing the necessary internal and external resources;
- ensuring that people are competent;
- explaining how people can help;
- managing your communications;
- controlling documented information;
- providing the infrastructure that your processes must have;
- providing the appropriate environment for your processes;
- providing suitable monitoring and measuring resources;
- providing information to assist process operations;
- managing your communications;
- controlling documented information;
- including all of the documented information that your QMS needs;
- managing the creation and revision of documented information;
- controlling the management and use of documented information;
- controlling your organisation's documents and records.

Clause 8: Operations

Operational processes:

- Develop, implement and control your operational processes.
- Clarify how:
  - product and service requirements will be managed;
  - communications with customers will be handled;
  - product and service requirements will be specified;
  - product and service requirements will be reviewed.

Design and development services:

- Develop products and services.
- Create a design and development process when necessary.
- Plan product and service design and development activities.
- Determine product, and service design and development inputs.
- Specify how design and development process will be controlled.
- Check how design and development outputs are produced.
- Review and control all design and development changes.
- Monitor and control externally provided products and services by:
  - confirming that external products and services meet requirements;
  - establishing controls for externally provided products and services;
  - discussing your organisation's requirements with external providers.

Manage and control production and service provision activities by:

- establishing controls for production and service provision;
- identifying process outputs and controlling their unique identity;
- protecting property owned by customers and external providers;
- preserving outputs during production and service provision;
- clarifying and complying with all post-delivery requirements;
- controlling changes for production and service provision;
- implementing arrangements to control product and service release;
- controlling nonconforming process outputs, products and services.

Clause 9 – Evaluation*

Monitor, measure, analyse and evaluate your QMS performance:

- Establish a process for monitoring, measuring, analysing and evaluating.
- Obtain customer information.
- Monitor customer satisfaction.
- Analyse and evaluate the results of monitoring and measurement.

Use internal audits to examine conformance and performance:

- Audit your QMS at planned intervals.
- Develop an internal audit programme for your organisation.

Review the suitability, adequacy, and effectiveness of your QMS:

- Plan and perform management reviews at planned intervals.
- Generate management review outputs and document results.

Clause 10: Improvement*

- Determine improvement opportunities and make improvements.
- Control nonconformities and take appropriate corrective action.
- Document your nonconformities and the actions that are taken.
- Enhance the suitability, adequacy and effectiveness of your QMS.

Author's Hint

Please refer to Annex A of this publication to see a précis of the actual requirements of ISO 9001:2015, together with an overview of the content of the various clauses and sub-clauses contained in the standard, and the likely documentation that will be required; a short resumé of how these would effect an organisation, and a cross-reference to the previous ISO 9001:2008 clauses is also provided.

5.9 WHAT ARE THE MAJOR CLAUSE CHANGES SEEN IN ISO 9001:2015?

As the latest edition of ISO 9001 has now been produced in accordance with Annex SL, with text and terminology that will become common to all other management system standards, there have been a number of necessary changes to the content and structure of the new ISO 9001:2015 (Table 5.2).

5.9.1 Are there any other significant changes?

- The wording of ISO 9001:2015 has been emphasised so that it influences the type and complexity of management system required, the needs and expectations of interested parties and the QMS and its processes.
- Top Management is now expected to demonstrate leadership and commitment and to engage, direct and support persons to contribute to the effectiveness of the QMS.
- The need for a management representative has been removed!

*This is a new clause introduced in ISO 9001:2015.

TABLE 5.2 Changes incorporated in ISO 9001:2015

ISO 9001: 2015 Clause	Change	Details of change, and impact
3	Terms and definitions	In accordance with Annex SL, common terms and definitions for all management systems are now to be found in ISO 9000:2005. ISO 9001:2015-specific terms and definitions will be found (and explained) in the standard itself.
4	Context of the organisation	Organisations are now required to identify any internal and external issues that may impact their Quality Management System's ability to deliver its intended results. They are also required to develop a methodology for understanding the needs and expectations of 'interested parties' (i.e. those individuals and organisations affected by the organisation's decisions or activities).
4.3	Quality Management System scope	Greater emphasis has been placed on the need for the organisation's QMS to take into consideration the internal and external issues identified by the context of the organisation.
4.3	Process approach requirement	The process approach is now a mandatory requirement of ISO 9001:2015.
5	Leadership	The title 'Management Responsibility' has now been replaced by 'Leadership', and in future Top Management will be required to be actively involved in all aspects of their Quality Management System. The role of the 'management representative' (usually referred to as the Quality Manager) no longer exists, as the 2015 version of ISO 9001 has embedded the QMS into routine business operations – as opposed to operating as an independent system in its own right with its own dedicated management structure.
6.1.1 and 6.1.2	Risks and opportunities	The term 'Preventive action' has now been replaced by 'actions to address risks and opportunities'. Organisations are now required to agree, consider and, where necessary, take action to address any risks or opportunities that may impact (either positively or negatively) on their QMS's ability to deliver its intended outcomes or that could have an effect on 'customer satisfaction'.

TABLE 5.2 continued

ISO 9001: 2015 Clause	Change	Details of change, and impact
7.5	Documented information	💣 This is probably the most contentious change made by ISO 9001:2015, and is viewed by many as a step in the wrong direction! However, after much discussion it has been decided that the requirement for an organisation to have a fully documented Quality Manual, documented (and sometimes mandatory) procedures and records will be removed and replaced with the term 'documented information', which is defined as 'information that the organisation will be required to keep, control and maintain'. The way an organisation decides how to maintain their 'documented information' is left open but, in doing so, ISO believes that this will provide organisations with a far more flexible way of running their business.
8.2	Products and services	The term 'product' has been replaced by 'products and services', which further emphasises that ISO 9001:2015 is applicable to all organisations whether they are designers, manufacturers, installers and/or end users.
8.4	Control of externally provided products and services	'Purchasing' has been replaced by 'control of externally provided products and services', in order to cover all forms of external provision whether purchased directly from a supplier, via an associate company or sub-contractor, or by any other means.
8.7	Non-conforming processes	The 'control of nonconforming products' now includes nonconforming processes as well as outputs and services.
	Annexes	ISO 9001:2015 now has three informative annexes: Annex A clarifies the new structure, terminology and concepts underpinning the standard. Annex B provides a set of Quality Management Principles that have been copyedited and updated from ISO 9004. Annex C provides details from the ISO 10,000 series concerning customer satisfaction (codes of conduct, handling complaints and disputes, and guidelines for monitoring and measuring).

- Much more emphasis has now been placed on planning and how to achieve quality objectives.
- ISO 9001:2015 recognises the increased availability and use of modern business practices, such as electronic systems and outsourcing.
- There is now more emphasis on products and services, external provision of products and services, and external providers.
- An additional sub-clause has been introduced to cover 'design and development of products and services'.
- 'Evaluation' is a new requirement in 9001:2015 and includes monitoring, measurement, analysis and evaluation, internal audit and management review.
- The requirement for a documented preventive action procedure has gone.
- Identification of risk and risk control is now a *requirement*.
- Organisations are now required to determine and maintain the knowledge they possess. (This includes not just knowledge held in documents or on IT systems, but also in people's minds.)
- The organisation *must* determine interested parties relevant to the QMS, and the requirements of these interested parties.
- The organisation must plan, implement and control the processes needed to meet requirements.

5.9.2 What about the revised duties of Top Management?

Top Management are now *required* to have a far greater involvement in the management of the system, particularly with respect to policy, organisational roles, responsibilities and authorities, communication, etc.

Top Management *must* have a greater hands-on involvement and be able to demonstrate their commitment to the management system, demonstrate leadership and commitment, and to engage, direct and support persons to contribute to the effectiveness of the QMS.

In a nutshell, Top Management must:

- establish, implement, maintain and improve their organisation's QMS (including the processes needed and their interactions) in accordance with the requirements of this standard;
- ensure the integration of the QMS's requirements into the organisation's business processes;
- ensure that the Quality Management System can achieve its intended outcome(s);
- ensure that the 'process approach' is maintained and is embedded in all of the organisation's activities;
- promote awareness of the 'process approach';
- prevent and/or reduce undesired effects;
- achieve improvement;
- be customer focused;
- support other relevant management roles to demonstrate their leadership as it applies to their areas of responsibility.

It is believed that with the new standard in place, organisations will find it easier to incorporate their existing management system into the Core Business Processes and, in doing so, gain greater business benefit.

If you would like to know a bit more about ISO 9001:2015's requirements, then please turn to Annex A where the structure of ISO 9001:2015 is explained in a little more detail and information is provided about the various clauses and elements contained in the standard's sections and subsections, together with the documentation that will probably be required. A cross-reference to the previous ISO 9001:2008 clauses is also provided.

Author's Note

As you can see, the introduction of Annex SL has not dramatically changed the requirements of ISO 9001, merely its structure so that it will match up with other management system standards.

But what are these 'other standards'?

We have already touched on a few of these, but as it is quite possible that you might be involved in one of these 'sons of ISO 9001' I have included a small chapter (Chapter 6) briefly explaining the ones that I know about at the time of writing this book.

What other standards are based on ISO 9001:2015?

CONTENTS

Author's Note

There are many other standards that replicate or are broadly based on ISO 9001:2015 and as it is quite possible that some of these 'sons of ISO 9001' might have an impact on your organisation, I have included the following small chapter about the background to the ones that I have come across at the time of writing this book.

During the last decade, reliance on the ISO 9000 series of quality standards has become a growing trend worldwide, with not just large multinationals seeking registration to ISO 9001, but also an increasing number of small and medium-sized Enterprises (SMEs).

Indeed, *small* US companies, although not as quick to jump on the bandwagon as their General Motors-type counterparts (owing to the perceived costs of registering a small business), are now seeing the benefit of working in compliance with ISO 9001. Indeed (according to a recent McGraw-Hill study of those companies registered since 2008) over 65% were able to recover their ISO 9000 implementation costs in three years or less. Consequently, the USA has now become one of the world leaders in ISO 9001:2015 SME registered companies.

ISO and ANSI have always worked closely together in producing interpretative standards for both sides of the Atlantic, and previous versions of ISO 9000 standards have been frequently used as the generic template for other industry *management* system standards. Currently, although there are still a number of these other national (both US and European) industry management standards available, they are all gradually being rewritten (using the same Annex SL structure, format and terminology, etc.) around the requirements and recommendations of ISO 9001:2015.

Author's Hint

It has to be said, however, that as much of the old ISO 9001:2008 standard has been preserved, it can be assumed that any revisions of these particular standards that will be only of a minor nature.

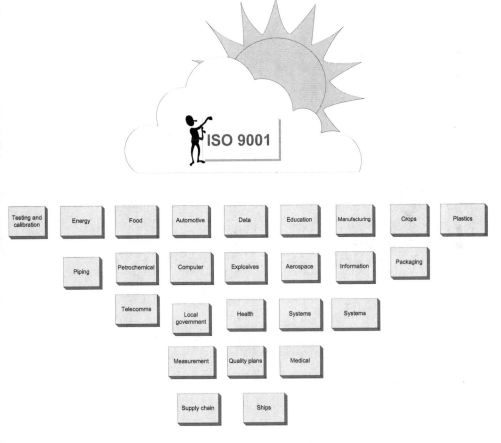

FIG. 6.1 The size of the ever-growing ISO 9001:2015 family

Listed below are some of the standards which have a 'relationship' with ISO 9001:2015 and of which I am aware at the time of writing this book.

TABLE 6.1 Other standards already based on ISO 9001:2008

Aerospace	AS 9100:2009
Auditing management systems	ISO 19011:2011
Automotive industry	ISO 16949:2009
Computer software	ISO 90003:2014
Crop production	ISO 22006:2009
Data	ISO/TS 8000–150:2011
Education	ISO 18091:2014
Electoral organisations	ISO 17582:2014
Energy managements systems	ISO 50001:2011
Explosive atmospheres	ISI/IEC 80079–34:2011
Food safety	ISO 22000:2005
Good manufacturing practice	ISO 15378:2011
Health care	IWA 1:2005
Human resources	ISO 76000:2015
Information security	ISO 27001:2013
Information technology	ISO 19796–1:2005
Local government	BS ISO 18091
Measurement manufacturing systems	ISO 10012:2003
Medical devices	ISO 13485:2012
Multi-layer piping systems	ISO/TS 21003–7:2008
Packaging: transport of dangerous goods	ISO 16106:2006
Petroleum, petrochemical and natural gas industries	ISO/TS 29001:2011
Quality Management Systems consultants	ISO 10019:2005
Quality Management Systems projects	ISO 10006:2003
Quality Plans	ISO 10005:2005
Ships and marine technology	ISO 30000:2009
Software engineering	ISO/IEC 90003:2014
Space systems	ISO 16192:2010
Supply chain management	ISO 28000:2007
Systems engineering	ISO TR 90005:2008
Telecommunications industry	TL 9000
Testing and calibration laboratories	ISO/IEC 17025:2005
Welding consumables	BS EN 12074:2000

6.1 Aerospace

AS/EN/JIS 9100: 2009 (*Quality Management Systems – Aerospace – Requirements*) is an international aerospace standard for Quality Assurance in design, development, production, installation and servicing of aircraft and aircraft systems.

6.2 Auditing management systems

ISO 19011:2011 (*Guidelines for auditing management systems*) provides the fundamental knowledge needed for the internal, external and third party audits of a management system. This has been expanded to reflect modern thinking and the intricacies of auditing *multiple* Management System Standards (MSS).

6.3 Automotive industry

ISO/TS 16949:2009 (*Quality Management Systems – Particular requirements for the application of ISO 9001:2008 for automotive production and relevant service part organisations*) defines the QMS requirements for automotive related services. When used in conjunction with ISO 9001:2008, it outlines best practice for design and development, production and installation when working with road vehicles.

6.4 Computer software

ISO/IEC 90003:2014 (*Software engineering – Guidelines for the application of ISO 9001:2008 to computer software*) provides guidance to organisations for the acquisition, supply, development, operation and maintenance of computer software.

Author's Hint
TickIT procedures relate directly to the requirements previously set out in ISO 9001:2008 and, similar to this standard, certification is conducted by an independent third party Certification Body using specialist auditors trained by the International Register of Certificated Auditors (IRCA) with the support of the British Computer Society.

6.5 Crop production

ISO 22006:2009 (*Quality Management Systems – Guidelines for the application of ISO 9001:2008 to crop production*) (as the title indicates) is aimed at assisting the crop industry in adapting the old ISO 9001:2008 processes. The term 'crop' includes seasonal crops (such as grains, pulses, oilseeds, spices, fruit and vegetables), row-planted crops that are cultivated, perennial crops that are managed over a period of time and wild crops that are not formally planted or managed. Horticultural crops provide an even greater range of types, from annual and perennial fruits, vegetables and ornamental flowering plants to perennial shrubs and trees, and root crops. These diverse crops require a broad range of planting, cultivating, pest control and harvesting methods and practices.

6.6 Data

ISO/TS 8000–150:2011 (*Data quality – Part 150: Master data: Quality management framework*) specifies the fundamental principles of quality data management and the requirements for implementation, data exchange and provenance. This standard also contains an informative framework identifying processes for data Quality Management which can be used in conjunction with, or independently of, Quality Management Systems standards, such as ISO 9001.

6.7 Education

ISO 18091:2014 (Quality Management Systems – Guidelines for the application of ISO 9001:2008) has been prepared so as to provide local governments worldwide with a consistent approach to Quality Management. It aims to 'translate' the technical language of ISO 9001:2008 into language that is more user-friendly for people who are involved in local government.

6.8 Electoral organisations

ISO 17582:2014 (Quality Management Systems. Particular requirements for the application of ISO 9001:2008 for electoral organisations at all levels of government) is, well, what it says 'on the tin'!

6.9 Energy management systems

ISO 50001:2011 (*Energy management systems – Requirements with guidance for use*) specifies requirements for establishing, implementing, maintaining and improving an energy management system.

6.10 Explosive atmospheres

ISO/IEC 80079–34:2011 (*Part 34: Application of quality systems for equipment manufacture*) specifies particular requirements and information for establishing and maintaining a quality system to manufacture Ex (explosive) equipment, including protective systems, in accordance with the Ex certificate.

6.11 Food safety

ISO 22000:2005 (*Food safety management systems – requirements for any organisation in the food chain*) specifies requirements for a food safety management system where an organisation in the food chain needs to demonstrate its ability to control food safety hazards in order to ensure that food is safe at the time of human consumption.

6.12 Good manufacturing practice

ISO 15378:2011 (*Primary packaging materials for medicinal products*) specifies requirements for a QMS where an organisation needs to demonstrate its ability to provide primary packaging materials for medicinal products which consistently meet customer requirements, including regulatory requirements and international standards applicable to primary packaging materials.

6.13 Health care

IWA 1:2005 (*Quality Management Systems – Guidelines for process improvements in health service organisations*) provides additional guidance for health service organisations involved in the management, delivery or administration of health service products and services.

6.14 Human resources

BS 76000:2015 (*Human resource. Valuing people. Management system. Requirements and guidance*) provides a framework for organisations to value people, for the mutual benefit of both parties.

6.15 Information technology

ISO 19796–1:2005 (*Information technology – Learning, education and training – Quality Management, Assurance and metrics – Part 1: General approach*) provides a framework to describe, compare, analyse and implement Quality Management and Quality Assurance approaches.

6.16 Information security

ISO 27001:2013 (*Information security management systems – Requirements*) specifies the requirements for establishing, implementing, operating, monitoring, reviewing, maintaining and improving a documented Information Security Management System (ISMS), taking into consideration the organisation's overall business risks.

6.17 Local government

BS ISO 18091 (*Quality management systems – Guidelines for the application of ISO 9001:2008 in local government*) provides local governments with guidelines for the voluntary application of ISO 9001 on an integral basis.

6.18 Measurement manufacturing systems

ISO 10012:2003 (*Measurement management systems*) specifies generic requirements and provides guidance for the management of measurement processes and
measuring equipment used to support and demonstrate compliance with
metrological requirements.

6.19 Medical devices

ISO 13485:2012 (*Medical devices – requirements for regulatory purposes*) specifies
the requirements for a QMS where an organisation needs to demonstrate its
ability to provide medical devices and related services that consistently meet
customer requirements and regulatory requirements applicable to medical devices
and related services.

As patient safety is involved, *all* of the requirements of ISO 13485:2012 are
mandatory!

6.20 Multi-layer piping systems

ISO/TS 21003-7:2008 (*Multi-layer piping systems for hot and cold water
installations inside buildings – Part 7: Guidance for the assessment of conformity*)
is applicable (in conjunction with the other parts of the ISO 21007 series) to hot
and cold water installations inside buildings for the conveyance of water – whether
or not the water is intended for human consumption (domestic systems) or for
heating systems – under specified design pressures and temperatures appropriate
to the class of application.

6.21 Packaging: transport packages for dangerous products

ISO 16106:2006 (*Packaging – Transport packages for dangerous products –
Dangerous products packagings, intermediate bulk containers and large packagings
– Guidelines for the application of ISO 9001*) provides guidance on quality
management provisions applicable to the manufacture, measuring and monitoring
of design type approved dangerous products packagings, intermediate bulk
containers (IBCs) and large packagings.

6.22 Petroleum, petrochemical and natural gas industries

ISO/TS 29001:2011 (*Petroleum, petrochemical and natural gas – Requirements for
product and service supply organisations*) defines the QMS for the petroleum,
petrochemical and natural gas industries.

6.23 Quality Management System consultants

ISO 10019:2005 (*Guidelines for the selection of Quality Management System
consultants and use of their services*) provides guidance on the factors to be taken
into consideration when selecting a Quality Management System consultant.

Author's Hint

This standard will be especially useful for SMEs who are or may be considering drawing up an ISO 9001:2015-compliant QMS with the assistance of an outside consultancy. It will also be of value to consulting organisations themselves.

6.24 Quality Management Systems projects

ISO 10006:2003 (*Guidelines for Quality Management in projects*) provides guidance on the application of Quality Management in projects of varying complexity, small or large, of short or long duration, in different environments, and irrespective of the kind of product or process involved.

6.25 Quality Plans

ISO 10005:2005 (*Guidelines for Quality Plans*) provides guidelines for the development, review, acceptance, application and revision of quality.

6.26 Ships and marine technology

ISO 30000:2009 (*Specifications for management systems for safe and environmentally sound ship recycling facilities*) provides requirements for the development and implementation of procedures, policies and objectives that will enable safe and environmentally sound ship recycling operations in accordance with national and international standards.

6.27 Software engineering

ISO/IEC 90003:2014 (*Software engineering – Guidelines for the application of ISO 9001:2008 to computer software*) provides guidance for organisations in the application of ISO 9001:2008 for the acquisition, supply, development, operation and maintenance of computer software and related support services.

6.28 Space systems

ISO 16192:2010 (*Space systems – Experience gained in space projects [Lessons learned]*) outlines principles and guidelines that are applicable to all space project activities (e.g. management, technical, quality, cost and schedule).

6.29 Supply chain management

ISO 28000:2007 (*Specification for security management systems for the supply chain*) specifies requirements for security management, including those aspects critical to the security assurance of the supply chain.

6.30 Systems engineering

ISO TR 90005:2008 (*Systems engineering – Guidelines for the application of ISO 9001 to system life cycle processes*) provides guidance for organisations involved in the application, acquisition, supply, development, operation and maintenance of systems and related support services.

6.31 Telecommunications industry

TL 9000 (*Quality Management standard for the telecommunication sector*) is a set of quality system requirements for the telecommunications industry which were originally developed by the QuEST Forum (Quality Excellence for Suppliers of Telecommunications Leadership). First published in November 1999, it was then updated to conform to the old ISO 9001:2008.

6.32 Testing and calibration laboratories

ISO/IEC 17025:2005 (*General requirements for the competence of testing and calibration laboratories*) specifies the general requirements for testing and calibration that is completed using standard methods, non-standard methods and laboratory-developed methods.

6.33 Welding consumables

BS EN 12074:2000 (*Quality requirements for manufacture, supply and distribution of consumables for welding and allied processes*) specifies tools for communication between a purchaser and a supplier of welding consumables within quality systems, such as those based upon ISO 9001.

Author's Note

Although this book has only provided you with a very brief overview of what ISO 9001:2015 is all about it should nevertheless be sufficient to enable you to produce your own Quality Management System for your organisation – if you really want to.

Assuming that is what you have done, in Chapter 7, we shall now look at how your new QMS can be improved and you can become an ISO 9001:2015 certified organisation.

But even if you don't want to go completely down that sort of road you can, with the information that you have so far received, 'work in compliance with ISO 9001:2015' which in many cases is sufficient to enable you to tender for some of the more demanding contracts.

Good luck!

Chapter 7

What to do once your QMS is established

CONTENTS

> ### Author's Note
>
> If you are a newcomer to ISO 9001 and Quality Management, I hope that this book has given you sufficient information to be able to keep up with 'quality experts' and understand what they are talking about. Perhaps this book has also given you an insight into the advantages that can be gained over your competitors by compiling a Quality Management System for your own company and even biting the bullet and becoming a certified ISO 9001:2015 company yourself?
>
> Having shown you what should be included in your Quality Management System, you will obviously want to know what the next steps should be. So in this final chapter I will show you how to conduct an internal audit and continually improve your QMS. I will also show you what would be involved if you decided that you wanted to be an ISO 9001:2015 company yourself.
>
> If your company is already certified to the 2008 version of this standard, then I have included a small section on how you can upgrade your existing system to meet the requirements of ISO 9001:2015.

7.1 INTERNAL AUDITS

Having set up your management system and implemented appropriate process metrics, you could be excused to think that this will provide you with sufficient confidence that everything within your business is functioning properly. This is not the case, as you will still have no idea where the strengths and weaknesses lie in your management system and this is where auditing can be used as a means of establishing whether:

- your management system is being used as it was originally intended;
- there are areas in the system which could be improved.

Author's Hint

In the previous edition of this standard, ISO required a mandatory procedure to be set up for auditing the organisation on a regular basis. Although this requirement has now been removed, I would nevertheless recommend that if you already have one in place that you continue to use it!

Auditing is all about checking that you conform to predetermined requirements (i.e. you do what you say you do!).

For example, if you state in a process that Staff must wear hard hats to perform a particularly dangerous job and a safety auditor visits a site only to discover they are all wearing flat caps, then your process is not being followed. The auditor will ensure that hard hats are worn, by issuing a nonconformance.

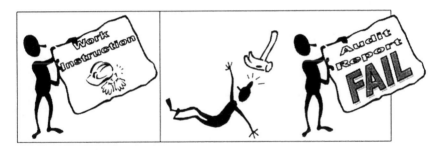

FIG. 7.1 Auditing is good for you!

This is a very typical example of how an auditor can maintain discipline within a workforce and ensures your management system is being adhered to. The auditor may also have saved someone's life, not to mention the costs of an accident claim.

Auditing generally follows a linear process, starting with establishing the criteria against which you are auditing and leading to a report concluding whether the criteria are being met. Should the audit find problems with the performance of a process, then you will implement corrective action aimed at preventing reoccurrence. A simple process map of the internal audit procedure is shown in Figure 7.2.

Remember, auditing is not a witch-hunt aimed at finding a fault and then blaming someone for it. When things go wrong it is usually the fault of the process, not the person performing it.

There are three generic categories of audit you may come across:

- **First party:** Also known as internal auditing, where (as the name suggests) members of an organisation look inwards at their own processes. This is the least effective form of auditing, as generally the auditors will find it difficult to criticise their own work. To minimise this, it is desirable to get Staff to audit each other's processes, thereby instilling a degree of independence.
- **Second Party:** These are audits carried out by your customers to satisfy themselves you are capable of doing a job and are generally referred to as vendor audits.
- **Third Party:** Personnel who are neither Staff working within your business nor your customers will carry out this type of audit. So who are they? Generally they are employees of accredited Certification Bodies. These are Notified Bodies who will audit your business and, if found to be compliant, certify you to ISO 9001. In addition to these three categories, other types of audit are available that can be used to measure conformance to ISO 9001:2015, such as:
  - **System Audits:** Carried out to ensure a business management system is sufficiently comprehensive to control all of the activities within that business. Generally, this type of audit would look for gaps in the

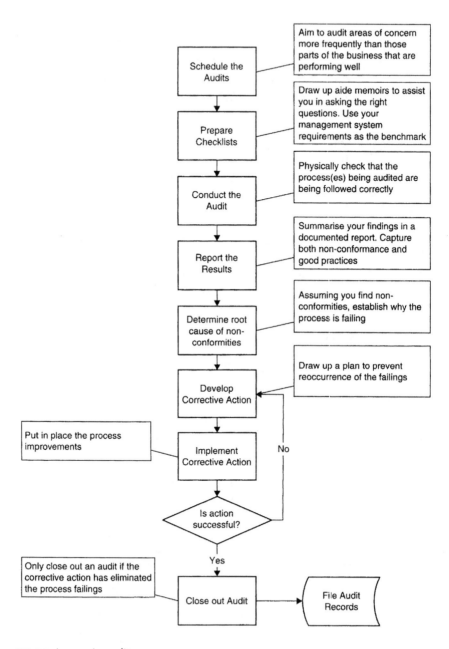

FIG. 7.2 A generic audit process

management system that may result them not achieving their business objectives.

- **Process Audits**: These focus specifically on single processes to verify whether they are capable of delivering the outputs expected of them.
- **Management Audits**: Checks carried out to see whether an organisation's strategic plan reflects their business objectives and, more specifically, that they have met the requirements of the intended market.

Author's Hint

These are just a few of the different audit types. If you would like to see examples of these audits plus a whole host of audit checklists and typical auditing forms, etc., may I suggest (and no, I haven't got my 'Marketing Hat' on, it is just that I want to help you!) that you get a copy of one of the other books that I have written in this ISO 9001 series, namely, 'ISO 9001:2015 Audit Procedures'.

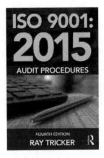

FIG. 7.3 Internal audits

7.2 CERTIFICATION

Many of today's contracts (particularly those for the military or government – but also for many large businesses) now *insist* that applicants hold a current ISO 9001:2015 certificate before they will even be considered for the job – whatever the deliverable, product or service.

Indeed, in many tender documents this is the first question that has to be answered, and to a small business (which has probably never envisaged going outside its own particular market) it can cause a lot of anguish.

If the organisation has a well-documented, properly audited, management-led QMS in place that is subject to continuous improvement and is always seeking customer satisfaction, then the road to gaining ISO 9001:2015 certification need not be too onerous. However, the obvious questions that a small business is going to ask are: 'How much is it going to cost to implement and operate?' And 'is it going to be worth having the ISO Certificate hanging on the Managing Director's wall?!!'

Obviously, each business is different and the cost of ISO 9001 registration will vary depending on the size and complexity of your organisation and on whether you already have some elements of a QMS in place. Consequently, although no book could possibly answer this question with any accuracy, in Table 7.1 (see section 7.2.5) I have shown typical budgetary costs for obtaining ISO 9001:2015 certification which will, I hope, be beneficial to you.

 But naturally the first question that has to be answered is: Do you actually need to become an ISO 9001 registered company or would simply *'working in compliance with ISO 9001:2015'* be sufficient? – and only your Top Management will know the answer to that one!

7.2.1 What are the benefits of achieving certification to ISO 9001:2015?

The benefits of gaining ISO 9001 certification by an accredited Certification Body include the following:

- shows commitment to quality, customers and a willingness to work towards improving efficiency;
- demonstrates the existence of an effective Quality Management System that satisfies the rigours of an independent, external audit;
- enhances company image in the eyes of customers, employees and shareholders alike;
- also gives a competitive edge to an organisation's marketing.

An organisation can also decide to seek certification because it:

- is a contractual or regulatory requirement;
- is necessary to meet customer preferences;

- falls within the context of a risk management programme; and
- helps motivate Staff by setting a clear goal for the development of its Management System.

Nevertheless, the whole aim of companies becoming accredited to the ISO 9001:2015 Standard is to ensure that their products and services are safe, reliable, of good quality and satisfy their customers' requirements.

For businesses, the Standards are strategic tools that lower costs by reducing waste and errors whilst at the same time increasing productivity. Accreditation helps companies not only to access new markets, but in doing so to become more professional as an organisation and improve their client relationships – as shown by the results of a recent study (i.e. 2015) of mixed businesses.

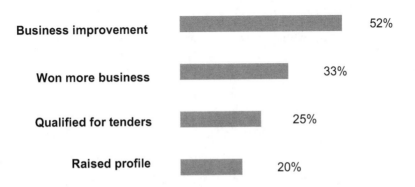

Business improvement   52%

Won more business   33%

Qualified for tenders   25%

Raised profile   20%

FIG. 7.4 The benefits of achieving ISO 9001

More than 80% confirmed that they had become more competitive and had won more business as a result of becoming ISO 9001:2015 certified and, in the last decade, certification to ISO 9001 has grown into a worldwide business requirement so that by the end of 2015, just within Europe, there were close to over 3/4 of a million companies certified to ISO 9001 – 10% of whom were from the UK alone!

7.2.2 What are the differences in being a certified, accredited and/or registered ISO 9001:2015 organisation?

- **Certification:** is the provision by an independent body of written assurance (a certificate) that the deliverable (product, service or system in question) meets specific requirements.
- **Registration:** Certification is very often referred to as registration in North America – but it is one and the same thing.
- **Accreditation:** This is the formal recognition by an independent accreditation body (e.g. in the UK, UKAS, and in the USA, ANAB) that a Certification Body has been formally approved as being capable of carrying out the certification of an organisation's QMS. Accreditation is not obligatory but it adds another

level of confidence, as 'accredited' means the Certification Body has been independently checked to make sure it operates according to international standards.

7.2.3 But can't I just work 'in compliance' with ISO 9001?

If your organisation simply needs to prove that it only has to work in compliance with the requirements of ISO 9001:2015, then all that is required is to possess an auditable and complete set of documents and records showing how you maintain overall quality in the Design, Build, and Operation & Maintenance (DBOM) of your products and services.

7.2.4 OK, so how do I become an ISO 9001 registered organisation then?

Assuming that you definitely decided that you want to be ISO 9001 certified then you will need to:

- Purchase a copy of ISO 9001:2015 (or get yourself a copy of 'ISO 9001:2015 for Small Businesses' (see Bibliography) which includes a generic QMS that you can customise to suit your own organisation).
- Identify which requirements of the ISO 9001:2015 standard are applicable to your type of organisation.
- Agree your organisational quality objectives, Quality Policy and Quality Plans.
- Produce a fully documented Quality Management System (consisting of a Quality Manual plus its associated Quality Processes, Procedures and Work Instructions) that is fully compliant with the requirements of ISO 9001:2015.
- Implement these processes and procedures throughout your organisation.
- Complete a series of internal audits to ensure that these procedures are suitable and adhered to.

TABLE 7.1 Examples of Accreditation Bodies

UKAS MANAGEMENT SYSTEMS 060	To find out who are certified within the UK you will need to contact the United Kingdom Accreditation Service (UKAS).
ANAB	Within the USA, the ANSI-ASQ National Accreditation Board ANAB) accredits Certification Bodies.

Reproduction of these logos by kind permission of the United Kingdom Accreditation Service and the ANSI-ASQ National Accreditation Board)

Once all the requirements of ISO 9001 have been met within your organisation and all non-compliances identified by your internal audit have been rectified, it is time for an external audit by a third party Certification Body.

Certificates are awarded by accredited Certification Bodies (also known as registrars) who have met the requirements of an accreditation body such as UKAS or ANAB (see Table 7.1).

Be warned, however: not *all* companies who profess to being able to award ISO 9001:2015 certificates are accredited!

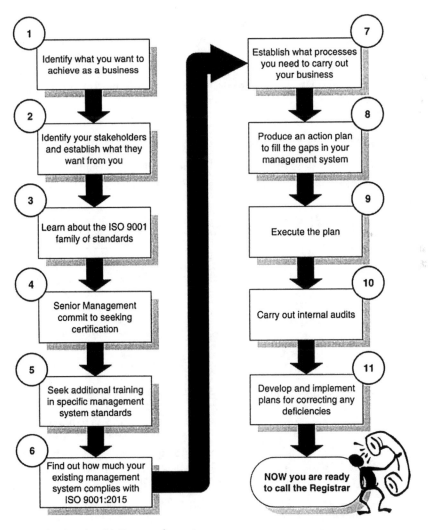

FIG. 7.5 Getting the QMS up and running

FIG. 7.6 Typical UKAS certification logo

Reproduction of these logos by kind permission of the United Kingdom Accreditation Service

The cost of certification can vary significantly as Certification Bodies have different pricing structures. Some will charge for each and every visit, assessment and follow-up surveillance inspections. Others may be happy to settle for a one-off fixed payment to take you through the certification process, followed by an annual renewal fee. When considering a suitable Certification Body you should obtain a number of quotes to establish the best offer.

7.2.5 What would the budgetary costs be for doing all this?

The cost for an organisation seeking registration in the UK would (at the time of publication) be in the region of the figures shown in Table 7.2.

TABLE 7.2 Budgetary costs for obtaining ISO 9001:2015 certification

Enterprise category	Head count	First stage Third party audit	Second stage Third party audit	Yearly assessments
Medium-sized	<250	£600	£2400	£1200
Small	<50	£600	£1200	£600–£1200
Micro	<10	£300	£600	£600

7.2.6 How long will it take to become certified?

With the right preparation and a good understanding of what is required for ISO 9001 certification, most organisations can expect to achieve certification within six to nine months depending on their size and complexity – but I cannot over-emphasise the importance of having *the complete backing* of Top Management, and it is absolutely vital that you have someone (either internal or perhaps in the case of a micro business, an external consultant) who has experience of implementing QMS and who knows what will be required in order to gain ISO 9001:2015 accreditation.

7.2.7 What is the certification process?

The chosen Certification Body will thoroughly review your quality processes and procedures, etc. to see that your organisation's management programme is measurable and achievable.

For a small business this could either consist of a desktop study or a one-day visit to your organisation's premises. For a large organisation, a site visit would probably be required – dependent on your product or service. The Certification Body may also send you a few simple questionnaires to complete.

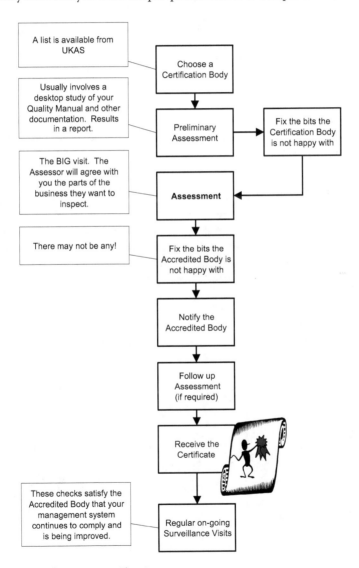

FIG. 7.7 A typical route to certification

Assuming that this Stage 1 audit is successful, then this will be followed at a later date by a full, on-site audit to ensure that working practices observe all of your policies, procedures and stated objectives and that appropriate records are maintained.

If this Stage 2 audit is successful, then the Certification Body will issue you with a certificate of registration to ISO 9001 and this will then be followed by annual (or in some cases, depending on the size and complexity of your organisation, bi-annual) surveillance visits to ensure that the system continues to work.

7.3 WHAT DO I DO ONCE MY QMS IS ESTABLISHED?

So you have finally managed to set up your Management System, got everyone working in accordance with it and are starting to reap the benefits of having a consistent approach to doing business. You could be excused to think all the hard work is over. This is far from the truth, as we already know because ISO 9001:2015 calls for you to continually improve the suitability, adequacy and effectiveness of your QMS.

So in the following sections we will look at the activities you should carry out once you have set up your QMS so that you can do exactly that – i.e. continually improve it.

7.3.1 Improvement

In Section 5.5 we learned that 'Improvement' is one of the seven principles of sound management practice. This principle requires a closer look, as any business that fails to develop will ultimately be left behind by the competition. Consider this: You may well have invented the hourglass, but there is always someone prepared to progress the design a little further. Resting on your laurels is not an option!

In the past, many companies who attained ISO 9001 certification did not seek to improve their QMS. After the initial enthusiasm of attaining the award – and the management had a certificate on the wall in reception – then a degree of apathy would set in and bad habits re-emerge. It was possible to ignore their QMS until

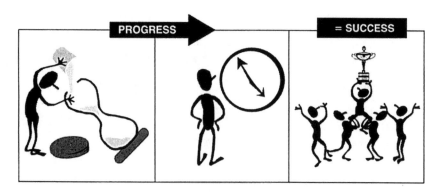

FIG. 7.8 Continual improvement keeps you ahead

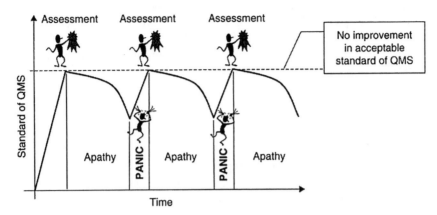

FIG. 7.9 The apathy–panic cycle!

the next assessment was due. This in turn resulted in a decline in standards between assessments, and a knee-jerk reaction and subsequent panic was then required to bring the system back up to an acceptable level.

This approach to retaining the certificate does not instil confidence in your QMS. Neither does it encourage you to seek ways in which to do things better. In short, your business processes become stagnant.

The only way in which a business will maintain viability and remain competitive is to continually refine the way in which it works as an ongoing activity. 'Continual improvement' avoids the apathy–panic cycle and aims to keep a business in business. It is a never-ending quest for perfection.

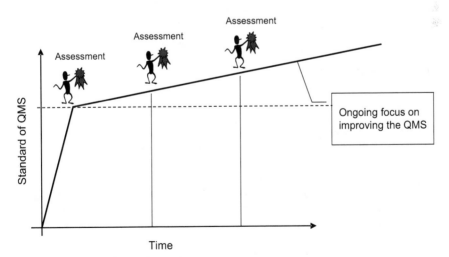

FIG. 7.10 The quest for perfection

Seeking utopia is all well and good, but it is not always easy to see where to find it. Your QMS may seem perfect for your purposes and improvements appear impossible to find. So where to start looking?

There is an old adage that says if you can't measure it, you can't manage it. This is especially true of a process-driven QMS. By measuring the performance of processes you will see how they are performing and whether they should be improved.

7.3.2 Process metrics

Just about every process can be measured to see how it is performing. Take for example the process of waking up. The purpose of this process is to wake someone up. So the obvious measure (or metric) would be the number of times the sleeping person responded to the alarm clock by getting out of bed.

Now supposing you set a benchmark or target to this metric, by saying an acceptable standard would be the man getting up on time for 80% of each month. By monitoring his actions over an extended period you would be able to establish whether this target is being met.

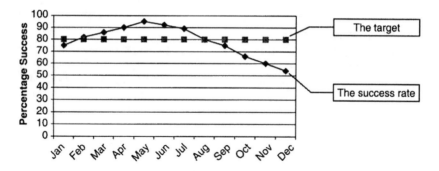

FIG. 7.11 The wake-up metric

In the above example, it can be seen that this target is *not* met during the winter months. This therefore represents an opportunity for process improvement. Maybe this would be through altering the process inputs, such as redefining the time he goes to bed, altering the bedroom temperature or the volume of the alarm. In this way the process is refined to better achieve its intended goal.

This is a very simple example but it demonstrates that processes can be measured and, if necessary, improved upon. Of course once a process is refined you must continue to monitor the metrics to establish whether the process has indeed been improved.

7.3.3 Process improvement tools

There are many tools available to help in targeting and implementing process improvements. The one still showing no signs of losing its popularity is known as Six Sigma, which (although not applicable to every QMS) is still relevant today. To be more precise, Six Sigma is a carefully packaged collection of tools (or toolbox) that can be applied as appropriate to the process being considered.

Six Sigma tools and techniques are numerous but all have one ultimate goal, that being the improvement of business processes.

FIG. 7.12 The Six Sigma tools of the trade

Six Sigma is a far-ranging subject about which numerous books have been written, and *ISO 9001:2015 in Brief* could not possibly do it justice. Suffice to say that it is a problem-solving methodology that focuses on the measurement of processes, subsequent identification and elimination or control of the root causes of defects/errors, and the institutionalisation of improvements within the management system.

For those of you with a mathematical bent, it aims to reduce defects to 3.4 in every 1 million – which equates to 0.0003%!

Six Sigma is frowned upon by some managers, however, as they believe that the tools are only developments of those already used by quality professionals since the early days of formal business management controls using statistical techniques (circa 1920). The tools have simply been better structured and used in a more systematic and logical sequence. Six Sigma can live quite happily in companies who already embrace ISO standards for business management. Indeed, it has potential to become the key contributor to the continual improvement culture required by ISO 9001.

Experience shows that companies who utilise Six Sigma as part of their continual improvement strategy not only become more efficient, but also have fewer problems attaining ISO 9001 certification.

7.3.4 How do I apply Six Sigma to improve my Processes?

Six Sigma uses the 'DMAIC' problem-solving model: Define, Measure, Analyse, Improve and Control. Each phase is summarised below:

- **Define:** This phase aims at pinpointing those processes that are causing your company problems. Management will also define the requirements for each process, i.e. set the standard to be expected.

- **Measure**: This phase gathers data to establish the current level of process performance.

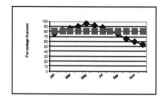

- **Analyse:** This is where the root cause of the problem is identified through the critical analysis of process metric data.

- **Improve:** Revolves around developing and testing ways in which the process can be improved.

- **Control:** Is simply putting in place whatever means are needed to keep the refined process in check. This could be through defining upper and lower tolerances, inspection requirements, methods of monitoring, etc.

7.3.5 Why have six Sigmas when one would do?

Again, for those of you with a mathematical bent, 'Sigma' is the value of the process standard deviation for a given characteristic.

In simpler terms, consider an archery target.

The 'given characteristic' would be the bullseye, because this is ultimately what you want to achieve. Fire a few arrows at the target and see how scattered they are from the bullseye. The greater the spread, the larger the standard deviation.

The term Sigma level refers to how many standard deviations fit between the mean (the target) and the specification limits (the edges of the target). Therefore, a process with a smaller standard deviation will be able to fit more Sigma levels into the same distance than a process with a large standard deviation. The higher the Sigma level, the better the quality of the product or service, i.e. the more arrows hitting the bullseye.

Table 7.3 displays various Sigma levels in terms of defects per million activities. An archer with a Sigma level of 6 would be an ideal Olympic champion, as only just over 3 of their 1,000,000 arrows would ever miss the bullseye!

TABLE 7.3 The 6th level is the best!

Sigma level	Defects per million	% Yield
1	691,463	30.854
2	308,538	69.146
3	66,807	93.319
4	6,210	99.379
5	233	99.977
6	3.4	99.99966

7.4 How do I upgrade my existing ISO 9001:2008 registration to meet ISO 9001:2015's requirements?

Author's Hint

Although organisations already registered to the 2008 standard will have up to three years following publication of the ISO 9001:2015 in which to re-certify, it is strongly recommended that if you are one of these organisations, you make a start on the transition to the new standard as soon as possible.

7.4.1 Does my organisation have to scrap its existing processes and procedures and start again?

No. Most organisations that have implemented ISO 9001: 2008 will find that their existing systems already meet many requirements of the new standard, and you can keep these in place with slight modifications in order to fully comply with ISO 9001:2015.

7.4.2 Does my organisation need to re-certify now that ISO 9001:2015 is published?

No. Organisations will have three years to make the move from ISO 9001:2008 to ISO 9001:2015.

7.4.3 How can I prepare for the new standard?

Strong leadership (i.e. by Top Management) has now become a major 'must' for all organisations, and these are now required to:

- be capable of demonstrating a thorough understanding of the business environment and how it impacts on their organisation's strategy. The system objectives need to be compatible with this strategy and set at relevant levels within the organisation, which will need to be able to demonstrate alignment between system objectives and its strategic direction;
- ensure that they have identified the significant risks that can have an impact on system objectives (e.g. customer satisfaction);
- ensure that there are clear responsibilities and authorities defined for all of the organisation's processes;
- review the organisation's internal communication channels for their effectiveness;
- review the process for managing change and development within the organisation so as to ensure that the effectiveness of the system is maintained during improvement and other organisational changes.

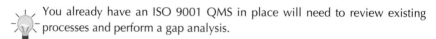 You already have an ISO 9001 QMS in place will need to review existing processes and perform a gap analysis.

7.5 How can all this be achieved?

Well, the simplest way is to make use of the following well-known, PDCA (Plan, Do, Check and Act) route shown in Fig. 7.14.

Note: Many of the leadership responsibilities are contained in the standard text of Annex SL. Therefore the principles and requirements for other management system standards will be very similar but with a changed focus to, for instance, environmental management or health and safety management.

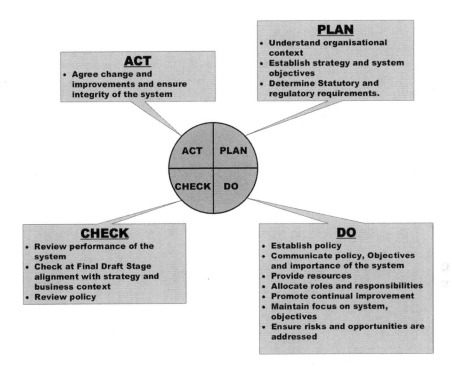

FIG. 7.13 How organisations should prepare for their change to ISO 9001:2015

 Author's Note

'ISO 9001:2015 in Brief' is only meant as a sort of quick reference to this standard of quality.

However, if you need a more in-depth book, then my main book in the current ISO 9001:2015 series (*ISO 9001:2015 for Small Businesses*, with its fully customisable generic QMS) is the book for you.

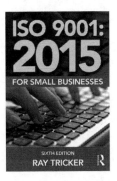

ISO 9001:2015 – A summary of requirements

CONTENTS

Author's Note

In this Annex, the structure of ISO 9001:2015 is explained in a little more detail and information is provided about the various clauses and elements contained in the standard's sections and subsections, together with the likely documentation that will be required. A cross-reference to the previous ISO 9001:2008 Clauses is also provided.

Please note that this is an extracted précis from another book in this series, *ISO 9001:2015 for Small Businesses*), which contains more in-depth information about ways of effectively complying with the requirements and recommendations of the standard.

Although some of the formal procedures contained in ISO 9001:2015 are mainly used by larger companies, there is absolutely no reason why smaller companies cannot adapt these procedures to suit their own purposes.

For example, ISO 9001:2015 Section 8.3 states that '*The organisation shall establish, implement and maintain a design and development process that is appropriate to ensure the subsequent provision of products and services*'.

Perhaps your company does not have a design office and this activity is achieved by an individual. But even though most of the requirements of Section 8.3 are probably inappropriate, the procedures required are still the same – so why not make use of them!

A detailed description of the main clauses making up ISO 9001:2015, together with explanations of how to meet the requirements of these clauses, now follows.

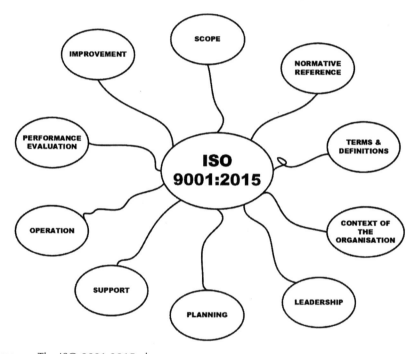

FIG. A1 The ISO 9001:2015 clauses

Please note that the structure and requirements featured in ISO 9001:2015 are generic with the intention that they can be applicable to any type of organisation, regardless of its type or size, or the products and services it provides.

1 SCOPE

ISO 9001:2015 Clause 1 'Scope'	This international standard specifies requirements for a Quality Management System when an organisation: • needs to demonstrate its ability to consistently provide products and services that meet customer and applicable statutory and regulatory requirements; • aims to enhance customer satisfaction through the effective application of the system, including processes for improvement of the system and the assurance of conformity to customer and applicable statutory and regulatory requirements
ISO 9001:2008 Clause	1 'Scope'
Proof	The intended outcomes of the Management System
Likely Documentation	Documented information containing everything related to Quality Control and Quality Assurance within an organisation.

2 NORMATIVE REFERENCES

ISO 9001:2015 Clause 2 'Normative references'	The following documents, in whole or in part, are normatively referenced in this document and are indispensable for its application ISO 9000:2015, 'Quality Management Systems – fundamentals and vocabulary'
ISO 9001:2008 Clause	2 'Normative Reference'
Proof	Reference standards or publications relevant to this particular standard
Likely Documentation	ISO 9000:2015 'Quality Management Systems – fundamentals and vocabulary'

3 TERMS AND DEFINITIONS

ISO 9001:2015 Clause 3 'Terms and definitions'	For the purposes of this document, the terms and definitions given in ISO 9000:2015 'Quality Management Systems – fundamentals and vocabulary' apply
ISO 9001:2008 Clause	3 'Terms and definitions'
Proof	That the relevant terms and definitions that apply to ISO 9000:2015 have been included in this standard
Likely Documentation	ISO 9000:2015 'Quality Management Systems – fundamentals and vocabulary'

4 CONTEXT OF THE ORGANISATION

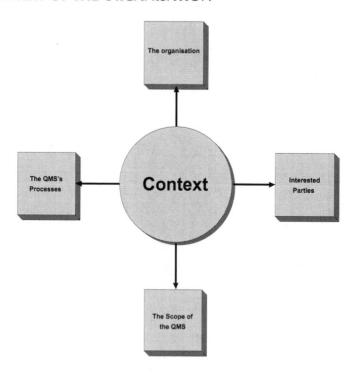

FIG. A2 Clause 4: Context

This clause is broken down into four separate sub-clauses which address the needs and expectations of interested parties and the scope of their QMS.

 Note: The list of interested parties that an organisation must consider should include: direct customers, end users, suppliers, subcontractors, distributors, retailers (or others involved in the supply chain) and regulators, etc.

4.1 Understanding the organisation and its context

ISO 9001:2015 Clause 4.1 'Understanding the organisation and its context'	'The organisation shall determine external and internal issues that are relevant to its purpose and its strategic direction and that affect its ability to achieve the intended result(s) of its Quality Management System'
Equivalent ISO 9001:2008 Clause	None (although in spirit this requirement was found under 9008 clause 1.1 'General')
Proof	An internal audit to determine positive and negative factors or conditions that need to be considered for the enhancement of their overall QMS
Likely Documentation	A formalised quality process

This is the first of two *new* clauses introduced into ISO 9001:2015 relating to 'Context' which requires organisations to demonstrate that they have a process that identifies, monitors, reviews and is capable of resolving any internal and external issues that are relevant to its purpose and which could have an effect on its ability to achieve the intended outcomes of their Quality Management System.

 Organisations will now need to provide evidence of this process to their auditors!

4.2 Understanding the needs and expectations of interested parties

This is another *new* clause introduced for ISO 9001:2015, which requires organisations to determine the boundaries and applicability of their QMS and to identify 'relevant interested parties'.

The organisation needs to make sure that they are completely aware of the overall (and specific) requirements of these interested parties – such as:

- direct customers;
- end users;
- suppliers, distributors, retailers or others involved in the supply chain; and
- regulators

who have the ability to impact (or potentially impact) the organisation's potential to consistently supply products and services that meet customer *and* applicable statutory and regulatory requirements.

ISO 9001:2015 Clause 4.2 'Understanding the needs and expectations of interested parties'	'The organisation shall determine: • the interested parties that are relevant to the Quality Management System; • the requirements of these interested parties that are relevant to the Quality Management System'
Equivalent ISO 9001:2008 Clause(s)	None (although in spirit this requirement was found under 2008 1.1 'General')
Proof	Speaking to the 'customer'
Likely Documentation	A formalised Quality Process

4.3 Determining the scope of the Quality Management System

Although the organisation has always been required to specify the scope of its QMS, this must now be done paying particular attention to the products and services it intends to supply.

ISO 9001:2015 Clause 4.3 'Determining the scope of the Quality Management System'	'The organisation shall determine the boundaries and applicability of the Quality Management System to establish its scope'
Equivalent ISO 9001:2008 Clause(s)	1.2 'Application' 4.2.2 'Quality manual'
Proof	A document which describes an organisation's quality policies, procedures and practices that makes up its QMS and which: • includes details of any associated documented processes and procedures; • indicates the sequence and interaction of these processes
Likely Documentation	A controlled document (such as a 'Quality Manual') containing everything related to Quality Control and Quality Assurance within an organisation

 This *new* clause replaces 'exclusions', which were previously referenced in ISO 9001:2008 as an acknowledgement that there might have been instances where it was impossible to apply a specific ISO 9001 requirement for a particular organisation.

4.4 Quality Management System and its processes

Note: The requirements contained in this clause are very similar to those found in the previous version of the standard, except that in the 2008 edition the requirements were less clear and were fragmented across a number of other clauses.

ISO 9001:2015 Clause 4.4 'Quality Management System and its processes'	'The organisation shall: • establish, implement, maintain and continually improve a Quality Management System, including the processes needed and their interactions • maintain and retain documented information to support the operation of its processes'
Equivalent ISO 9001:2008 Clause(s)	4 'Quality Management System' 4.1 'General requirements'
Proof	A definition of the processes necessary to ensure that a product conforms to customer requirements which are capable of being implemented, maintained and improved
Likely Documentation	Documented information (e.g. the Quality Manual or similar document)

The overall control of any organisation will rely on a number of different management disciplines, the most important of which is *Quality Management*. As this is the core of all organisational structures, the activities and processes that affect performance improvement will need to be described and defined by management. They will also need to ensure that these are clearly understood by the whole workforce, monitored (i.e. to evaluate improvement on a continuing basis) and managed.

 Where an organisation chooses to outsource any process that affects product conformity to requirements, the organisation will have to ensure that they have *full control* over such processes.

5 LEADERSHIP

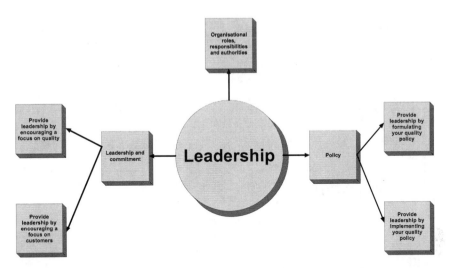

FIG. A3 Clause 5: leadership

In the 2015 edition of ISO 9001, 'management responsibility' is replaced by 'leadership'. Although 'leadership' would seem at first glance to be just a reiteration of what's gone before regarding policy, organisational roles, responsibilities, authorities and so on, in the 2015 edition there is now much more emphasis on this being seen as a 'hands-on' leadership as opposed to just 'management'.

5.1 Leadership and commitment

In the previous 2008 version of this standard, Top Management were required to 'establish a Quality Policy and to ensure that it was reviewed for continuing suitability'. ISO 9001:2015 now goes one step further and specifically *requires* that the Top Management 'establish, review and *maintain*' a Quality Policy.

5.1.1 General

ISO 9001:2015 Clause 5.1.1 'General'	'Top Management shall demonstrate leadership and commitment with respect to the Quality Management System'
Equivalent ISO 9001:2008 Clause(s)	5.1 'Management commitment'
Proof	A written demonstration of an organisation's commitment to sustainable Quality Management Auditable proof that all of the customer's requirements are (and have been) fully met
Likely Documentation	A Quality Manual containing: • a high-level policy statement concerning organisational objectives and quality policies; • a list of government regulatory, legal and customer-specific requirements; • procedures describing (amongst other topics) resource management, contract review, management review, financial business plan(s), etc.

Top Management now have a *mandatory requirement* to demonstrate leadership and commitment with respect to their organisation's QMS, and to ensure that the customer's (plus any applicable statutory and regulatory) requirements are agreed upon, understood and consistently met. They should try to create an environment in which people are fully involved and in which *their* QMS can operate effectively.

5.1.2 Customer focus

ISO 9001:2015 has now made it a *mandatory requirement* for Top Management to guarantee that their organisation consistently provides products and services that conform to customer requirements, that meet applicable statutory and regulatory requirements, and (of prime importance!) that enhance customer satisfaction!

ISO 9001:2015 Clause 5.1.2 'Customer focus'	'Top Management shall demonstrate leadership and commitment with respect to customer focus'
Equivalent ISO 9001:2008 Clause(s)	5.2 'Customer focus'
Proof	Auditable proof that all of the customer's requirements are (and have been) fully met
Likely Documentation	Procedures describing: • resource management; • contract review procedures; • management reviews; • financial business plan(s)

5.2 Policy

Whilst this is largely unchanged from the ISO 9001:2008 version of Quality Policy, it now has some additional requirements about its availability to interested parties with regard to documented information.

Quality Policies and quality objectives, therefore, need to be established in order to provide a general focus for the organisation. Policies and objectives determine the intended results and assist the organisation in applying their resources to achieve these results, and so these need to be established by Top Management in order to provide a general focus for the organisation.

 It is essential that quality policies are always interlinked with quality objectives.

5.2.1 Establishing the Quality Policy

ISO 9001:2015 Clause 5.2.1 'Establishing the Quality Policy'	'Top Management shall establish, implement and maintain a Quality Policy that: • is appropriate to the purpose and context of the organisation and supports its strategic direction; • provides a framework for setting quality objectives; • includes a commitment to satisfy applicable requirements; • includes a commitment to continual improvement of the Quality Management System'
Equivalent ISO 9001:2008 Clause(s)	5.3 'Quality policy'
Proof	A description of how an organisation approaches quality and how it ensures that this approach is appropriate for both the customer and its own needs
Likely Documentation	Documented information (usually in the form of a Management Manual) containing a high-level managerial statement concerning the organisation's Quality Policy, which provides details concerning the responsibilities, training and resources required for each organisational activity

5.2.2 Communicating the Quality Policy

ISO 9001:2015 now makes it a mandatory requirement for Top Management to ensure that their organisation consistently provides products and services that conform to their customer requirements and that meet all applicable statutory and regulatory requirements – this in addition, of course, to maintaining a focus on enhancing customer satisfaction!

ISO 9001:2015 Clause 5.2.2 'Communicating the Quality Policy'	'Top Management shall establish, implement and maintain a Quality Policy that: • is available and maintained as documented information; • is communicated, understood and applied within the organisation; • is available to relevant interested parties, as appropriate'
Equivalent ISO 9001:2008 Clause(s)	5.3 'Quality Policy'
Proof	Availability of documented information on how an organisation approaches quality and how it ensures that this approach is appropriate for both the customer and its own needs
Likely Documentation	A high-level managerial statement on an organisation's Quality Policy containing the clear responsibilities, training and resources required for each organisational activity

5.3 ORGANISATIONAL ROLES, RESPONSIBILITIES AND AUTHORITIES

This clause requires Top Management to assign QMS roles, responsibilities and authority to people who can be trusted to ensure that the organisation's QMS is capable of covering and meeting all of the requirements from ISO 9001:2015 that are necessary for their type of business.

ISO 9001:2015 Clause 5.3 'Organisational roles, responsibilities and authority'	'Top Management shall ensure that the responsibilities and authorities for relevant roles are assigned, communicated and understood within the organisation'
Equivalent ISO 9001:2008 Clause(s)	5.5.1 ('Responsibility and authority') 5.5.2 ('Management representative')
Proof	Documented information defining the roles, responsibilities, lines of authority, reporting and communication relevant to quality
Likely Documentation	• job descriptions and responsibilities; and/or • organisation charts showing lines of communication.

6 PLANNING

Clause 6 is all about how the organisation will prevent, or reduce, undesired effects (i.e. risks) and how it will ensure that it can still achieve the aims of its QMS whilst ensuring continual improvement.

Planning needs to include all the 'hows', 'whens', 'by whoms' and 'by whats' of possible risks occurring.

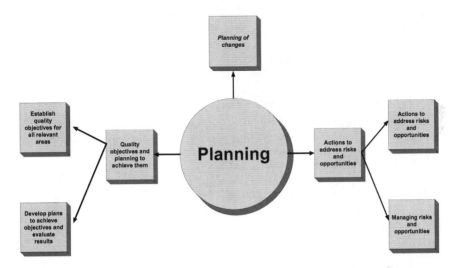

FIG. A4 Clause 6: planning

Risk-based thinking (with preventative action) is probably *the* most important requirement for the new version of ISO 9001, and once an organisation has highlighted the risks and opportunities in its products or services, it then needs to determine how and when these risks have to be addressed and, most importantly, by whom.

This is a *new* clause which will, in turn, reduce the need for corrective actions later on.

6.1 Actions to address risks and opportunities

The organisation should use preventive methodologies such as risk analysis, trend analysis, statistical process control, fault tree analysis, failure modes and effects and critical analysis to identify the causes of potential nonconformances.

ISO 9001:2015 Clause 6.1 'Actions to address risks and opportunities' (Part 6.1.1)	'When planning the Quality Management System, the organisation shall consider the issues referred to in 4.1 (i.e. concerning the organisation and its context) and the requirements referred to in 4.2 (i.e. concerning the needs and expectations of interested parties) and determine the risks and opportunities that need to be addressed'
Equivalent ISO 9001:2008 Clause(s)	None directly, although this new requirement extracts ideas previously found in 8.5.3 ('Preventative action'), 5.4.2 ('Quality Management System planning') and 7.1 ('Planning of product realisation')
Proof	A Project Risk Register containing documented information confirming that the organisation has identified (or has a process in place to identify) the possibility of risks occurring during the manufacture, supply, installation, usage and maintenance of a product or service
Likely Documentation	Processes and procedures used by Top Management – for example a SWOT (Strengths/Weaknesses/Opportunities/Threats) analysis – to: • give assurance that the QMS can achieve its intended results; • prevent or reduce undesired effects; and • achieve continual improvement.

'Risk analysis' is a *new* requirement (brought about by the introduction of Annex SL) which requires organisations to adopt a risk-based approach when planning the workflow of their business, and which means that they will have to decide those risks and opportunities that will have the potential to impact the operation and performance of their Quality Management System both positively and negatively.

ISO 9001:2015 Clause 6.1 'Quality objectives and how to achieve them' (Part 6.1.2)	'The organisation shall plan: <ul><li>actions to address these risks and opportunities;</li><li>how to 1) integrate and implement the actions into its Quality Management System processes (see 4.4); and 2) evaluate the effectiveness of these actions'</li></ul>
Equivalent ISO 9001:2008 Clause(s)	None directly, although this new requirement extracts ideas previously found in 8.5.3 ('Preventative action'), 5.4.2 ('Quality Management System planning') and 7.1 ('Planning of product realisation')
Proof	The availability of a regular ongoing and thorough internal auditing programme
Likely Documentation	Quality Process or Procedure for risk analysis

The organisation will need to:

- ensure that it has a strong risk management methodology in place;
- determine whether there are any risks and opportunities that could influence the performance of its QMS or disrupt or damage its operation;
- decide what actions should be taken to address any risks found and opportunities identified;
- ensure that it has proven risk treatment options available.

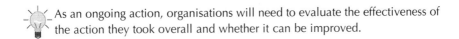 As an ongoing action, organisations will need to evaluate the effectiveness of the action they took overall and whether it can be improved.

6.2 Quality objectives and planning to achieve them

Top Management are required to document a set of quality objectives that the organisation must meet, particularly with respect to the conformity of products and services and the enhancement of customer satisfaction.

ISO 9001:2015 Clause 6.2 'Quality objectives and planning to achieve them' (Part 6.2.1)	'The organisation shall establish quality objectives at relevant functions, levels and processes needed for the Quality Management System' which shall be: • consistent with the Quality Policy; • measurable; • take into account applicable requirements; • relevant to conformity of products and services and to enhancement of customer satisfaction; • monitored; • communicated; • updated as appropriate.
Equivalent ISO 9001:2008 Clause(s)	5.4.1 'Quality objectives'
Proof	Quality objectives that Top Management expect to achieve within each function and level of the organisation
Likely Documentation	Policy statements defining the objectives of the company and those responsible for achieving these objectives

The development of how to plan and develop quality objectives has, with the introduction of ISO 9001:2015, now become more of a requirement than a recommendation and organisations now need to decide exactly what work is required in order to realise:

- its quality objectives;
- the resources and infrastructure required to complete this work;
- who will be responsible for this activity; and
- how, by whom and where the results are going to be evaluated and stored as documented information.

ISO 9001:2015 Clause 6.2 'Quality objectives and planning to achieve them' (Part 6.2.2)	'When planning how to achieve its quality objectives, the organisation shall determine: <ul><li>what will be done;</li><li>what resources will be required;</li><li>who will be responsible;</li><li>when it will be completed;</li><li>how the results will be evaluated'</li></ul>
Equivalent ISO 9001:2008 Clause(s)	5.4.1 'Quality objectives'
Proof	Quality objectives that Top Management expect to achieve within each function and level of the organisation
Likely Documentation	Policy statements defining the objectives of the company and those responsible for achieving these objectives

These quality objectives need to be firmly established during the planning stage and then circulated to all personnel involved so that they can readily translate them into individual (and achievable) contributions.

6.3 Planning of changes

When there is a need to make a change to the QMS then this must be completed in a controlled manner and any changes proposed (i.e. to processes, resources, responsibilities, methodologies, procedures, etc.) must be thoroughly reviewed and agreed by Top Management.

ISO 9001:2015 Clause 6.3 'Planning of changes'	'When the organisation determines the need for changes to the Quality Management System, the changes shall be carried out in a planned manner'
Equivalent ISO 9001:2008 Clause(s)	5.4.2 'Quality Management System planning'
Proof	The identification and planning of activities and resources required to meet an organisation's quality objectives: • current and future requirements; • the markets served; • the output from previous management reviews; • current product and process performance.
Likely Documentation	Processes and procedures developed by Top Management to define and plan the way that their organisation is run.

7 SUPPORT

Having thought about their overall organisational aims and policies, commitment and planning, organisations will now have to consider what resources are required to meet their goals and objectives.

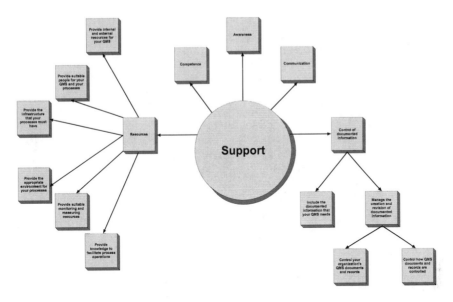

FIG. A.5 Clause 7: Support

7.1 Resources

7.1.1 General

ISO 9001:2015 Clause 7.1.1 'General'	'The organisation shall determine and provide the resources needed for the establishment, implementation, maintenance and continual improvement of the Quality Management System'
Equivalent ISO 9001:2008 Clause(s)	6 ('Resource management') and 6.1 ('Provision of resources')
Proof	An organisation's resources are identified with regard to people, materials, equipment (and infrastructure) training, induction, responsibilities, working environment, equipment needs, maintenance, etc.
Likely Documentation	• Quality Plans and Processes; • Quality Procedures; • Work Instructions.

The organisation needs to identify and make available all the resources (e.g. information, infrastructure, people, work environment, finance, support, etc.) required to implement and improve its QMS and its associated Quality Processes.

7.1.2 People

ISO 9001:2015 Clause 7.1.2 'People'	'The organisation shall determine and provide the persons necessary for the effective implementation of its Quality Management System and for the operation and control of its processes'
Equivalent ISO 9001:2008 Clause(s)	6.1 ('Provision of resources') and 6.2 ('Human resources')
Proof	How human resources to implement and improve the QMS are identified.
Likely Documentation	• Quality Plans; • Quality Procedures; • Work Instructions.

Human resources are the principal method of achieving product completion and customer satisfaction. The old adage 'a happy worker is a good worker' still stands true in this age of information technology, and with the increased training and education opportunities currently available, highly motivated, well-qualified personnel are at a premium.

To employ and retain the right sort of person for the job, management must determine the resources required, adequately define their responsibilities and authorities, establish their individual and team objectives and encourage recognition and reward.

They must also:

- consider career planning and on-the-job training (OJT);
- encourage innovation and effective teamwork;
- make full use of all available information technology;
- measure people's satisfaction.

The organisation is responsible for ensuring that all personnel are trained and experienced to the extent necessary to undertake their assigned activities and responsibilities effectively. Thus, whenever training needs have been identified, Top Management should endeavour to make the relevant training available and full records must be maintained of all training undertaken by employees.

7.1.3 Infrastructure

 Note: Other than a little bit of 'word-smithing' and minor revisions, this requirement is similar to the 2008 edition of the standard, in that the necessities for infrastructure are chiefly centred on identifying, providing and maintaining the means to enable the organisation's processes to operate effectively.

ISO 9001:2015 Clause 7.1.3 'Infrastructure'	'The organisation shall determine, provide and maintain the infrastructure necessary for the operation of its processes and to achieve conformity of products and services'
Equivalent ISO 9001:2008 Clause(s)	6.3 'Infrastructure'
Proof	How an organisation defines, provides and maintains the infrastructure requirements to ensure product conformity.
Likely Documentation	• policies, procedures and regulatory documents stating organisation and customer requirements; • budget and financial documents; • maintenance plans; • project plans identifying the human resources required to complete the task.

Infrastructure can include:

- buildings and associated utilities;
- equipment, including hardware and software;
- transportation resources;
- information and communication technology.

7.1.4 Environment for the operation of processes

Other than the need for a suitable work environment to be identified, provided and maintained (as opposed to just being 'determined and managed'), this requirement is similar to the previous edition of the standard.

ISO 9001:2015 Clause 7.1.4 'Environment for the operation of processes'	'The organisation shall determine, provide and maintain the environment necessary for the operation of its processes and to achieve conformity of products and services'
Equivalent ISO 9001:2008 Clause(s)	6.4 'Work environment'
Proof	How an organisation defines and organises its work environment.
Likely Documentation	• environmental procedures; • project plans; • budgetary processes; • legal processes and procedures.

An organisation's work environment is a combination of human factors (e.g. work methodologies, achievement and involvement opportunities, health and safety rules and guidance, ergonomics, etc.) and physical factors (e.g. heat, hygiene, vibration, noise, humidity, pollution, light, cleanliness and air flow). All of these factors influence motivation, satisfaction and performance of people and as they have the potential for enhancing the performance of the organisation, they must be taken into consideration by that organisation when evaluating product conformance and achievement.

7.1.5 Monitoring and measuring resources

General

Where an organisation uses monitoring or measuring to demonstrate that its products and services conform to requirements, it must make sure that it provides the necessary resources to ensure that its monitoring and measuring results are valid.

ISO 9001:2015 Clause 7.1.5.1 'Monitoring and measuring resources – General'	**'The organisation shall determine and provide the resources needed to ensure valid and reliable results when monitoring or measuring is used to verify the conformity of products and services to requirements'**
Equivalent ISO 9001:2008 Clause(s)	7.6 ('Control of monitoring and measuring equipment')
Proof	The controls that an organisation has in place to ensure that equipment (including software) used for proving conformance to specified requirements is properly maintained.
Likely Documentation	• equipment records of maintenance and calibration; • Work Instructions.

Measurement traceability

Without exception, all measuring instruments can be subject to damage, deterioration or just general wear and tear when they are in regular use in workshops and factories. The organisation's QMS should take account of this fact and ensure that *all* test equipment is regularly calibrated against a known working standard held by the manufacturer.

ISO 9001:2015 Clause 7.1.5.2 'Measurement traceability'	**'When measurement traceability is a requirement, or is considered by the organisation to be an essential part of providing confidence in the validity of measurement results, measuring equipment shall be:** • **calibrated or verified, or both, prior to use** • **identified in order to determine its status;** • **safeguarded from adjustments, damage or deterioration'**
Equivalent ISO 9001:2008 Clause(s)	7.6 'Control of monitoring and measuring equipment'
Proof	The controls that an organisation has in place to ensure that equipment (including software) used for proving conformance to specified requirements is properly maintained.
Likely Documentation	• equipment records of maintenance and calibration; • Work Instructions.

7.1.6 Organisational knowledge

This is a *new* requirement aimed at ensuring that organisations have in place a system for capturing and preserving knowledge and learning with regard to both the product and the organisation's QMS.

ISO 9001:2015 Clause 7.1.6 'Organisational knowledge'	'The organisation shall determine the knowledge necessary for the operation of its processes and to achieve conformity of products and services'
Equivalent ISO 9001:2008 Clause(s)	None
Proof	Knowledge that has become an integral part of the organisation, and has been maintained and made available on an as-required basis.
Likely Documentation	Documented information

7.2 Competence

Personnel performing work affecting product quality must be competent based on their education, training, skills and previous experience, and be able to perform their tasks with a minimum amount of supervision.

ISO 9001:2015 Clause 7.2 'Competence'	'The organisation shall determine the necessary competence of person(s) doing work under its control that affects the performance and effectiveness of the Quality Management System'
Equivalent ISO 9001:2008 Clause(s)	6.2.1 ('General') and 6.2.2 ('Competence, training and awareness')
Proof	Established procedures for: • the assignment of personnel; • training, awareness and competency.
Likely Documentation	• Quality Plans; • Quality Procedures; • Work Instructions.

7.3 Awareness

Clause 7.3 now makes it a *requirement* for people completing work under the organisation's control (*including* subcontractors) to be aware of the organisation's Quality Policy, any quality objectives that are relevant to them, how they will be contributing to the effectiveness of the QMS and what the implications would be of them not conforming to the organisation's QMS.

ISO 9001:2015 Clause 7.3 'Awareness'	'The organisation shall ensure that persons doing work under the organisation's control are aware of: • the Quality Policy; • relevant quality objectives; • their contribution to the effectiveness of the Quality Management System; and • the implications of not conforming with the Quality Management System requirements'
Equivalent ISO 9001:2008 Clause(s)	6.2.2 'Competence, training and awareness'
Proof	Established policies, processes and procedures
Likely Documentation	Quality Manual

Most organisations will recruit employees who are already well qualified and quite capable of meeting the relevant technical, skill, experience and educational requirements of the organisation. There will still, however, be a need for some additional system or contract-specific training, and all Staff have a responsibility for identifying and recommending the training needs of others and for ensuring that all employees allocated specific tasks are suitably qualified and experienced to execute those tasks.

7.4 Communication

The organisation is now required to decide on what it will communicate, when it will communicate, with whom it will communicate and how it will communicate.

ISO 9001:2015 Clause 7.4 'Communication'	'The organisation shall determine the internal and external communications relevant to the Quality Management System'
Equivalent ISO 9001:2008 Clause(s)	5.5.3 'Internal communication'
Proof	Confirmation that the requirements of an organisation's QMS are communicated throughout the company.
Likely Documentation	• notice boards; • in-house journals/magazines; • audio-visual; • e-information. Also: • team briefings; • organisational meetings.

7.5 Documented information

Author's Hint
From experience, the extent of documented information for a QMS can differ from one organisation to another, according to:

- the size of organisation and its type of activities, processes, products and services;
- the complexity of processes and their interactions;
- the competence of persons.

It can just be a few pages together with a few explanatory flowcharts – or a shelf-load of folders!

7.5.1 General

The requirement for the organisation to retain documented information for the effective operation of its QMS is unchanged from that included in ISO 9001:2008, Clause 4.2.

ISO 9001:2015 Clause 7.5.1 'General'	'The organisation's Quality Management System shall include: • documented information required by this international standard; • documented information determined by the organisation as being necessary for the effectiveness of the Quality Management System'
Equivalent ISO 9001:2008 Clause(s)	4.2 ('Documentation requirements') and 4.2.1 ('General')
Proof	Documented information that is usually found in a Quality Manual.
Likely Documentation	• document control procedures; • Work Instructions.

 With the increasing practice of organisations maintaining documented information via some electronic format, there will now be the need for an organisation to consider access controls (i.e. passwords/logins), authorisation levels and how the integrity of their documented information is maintained.

7.5.2 Creating and updating

The requirement for the organisation to create and manage their documented information is unchanged from that included in ISO 9001:2008.

ISO 9001:2015 Clause 7.5.2 'Creating and updating'	'When creating and updating documented information, the organisation shall ensure that it includes all the relevant information concerning: • description and identification (e.g. a title, date, author or reference number); • format (e.g. language, software version, graphics); • media (e.g. paper, electronic); • how it should be reviewed and approved for suitability and adequacy'
Equivalent ISO 9001:2008 Clause(s)	4.2.3 ('Control of documents') and 4.2.4 ('Control of records')
Proof	Documented information
Likely Documentation	Quality Process Quality Procedure

7.5.3 Control of documented information

The distribution of standard documents should be controlled and recorded on distribution lists, which also show the current issue status.

ISO 9001:2015 Clause 7.5.3 'Control of documented information'	'Documented information required by the Quality Management System and by this international standard shall be controlled to ensure that: • it is available and suitable for use, where and when it is needed; • it is adequately protected (e.g. from loss of confidentiality, improper use or loss of integrity)'
Equivalent ISO 9001:2008 Clause(s)	4.2.3 ('Control of documents') and 4.2.4 ('Control of records')
Proof	Documented information
Likely Documentation	Quality Process Quality Procedure

A master list of all documents should be maintained which clearly shows the current status of each document. This list needs to be available at all locations where operations effective to the functioning of the organisation's QMS quality are performed, and this distribution list needs to be reviewed and updated as changes occur and all invalid and/or obsolete documents/data must be immediately removed.

8 OPERATION

 Note: Clause 8 contains the lion's share of management system requirements addressing:

- in-house and outsourced processes;
- overall process management;
- necessary criteria to control these processes; and
- ways to manage planned and unintended change.

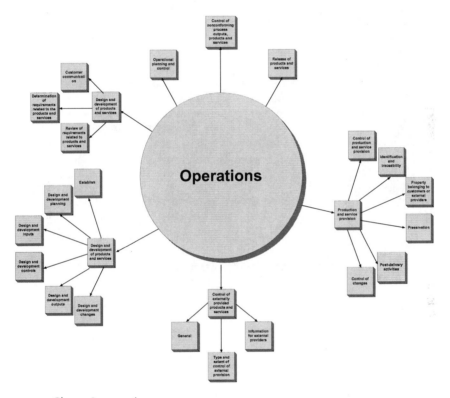

FIG. A6 Clause 8: operations

8.1 Operational planning and control

ISO 9001:2015 Clause 8.1 'Operational planning and control'	'The organisation shall plan, implement and control the processes needed to meet the requirements for the provision of products and services'
Equivalent ISO 9001:2008 Clause(s)	7 ('Product realisation') and 7.1 ('Planning of product realisation')
Proof	The availability of documented plans for processes that are required to realise a product – and the sequences in which they occur.
Likely Documentation	• process models (flow charts) showing the sequence of activities that an organisation adopts in order to produce a product; • Quality Plans; • documented QPs and WIs to ensure that Staff work in accordance with stipulated requirements; • documented information that proves the results of process control.

8.2 Requirements for products and services

8.2.1 Customer communication

Lines of communication between the customer and the organisation with regard to product information, enquiries and contract amendments, feedback (including complaints, etc.) must be clearly defined.

ISO 9001:2015 Clause 8.2.1 'Customer communication'	'Communication with customers shall include: • providing information relating to products and services; • handling enquiries, contracts or orders, including changes; • obtaining customer feedback relating to products and services, including customer complaints; • handling or controlling customer property; • establishing specific requirements for contingency actions, when relevant'
Equivalent ISO 9001:2008 Clause(s)	7.2 ('Customer-related processes') and 7.2.3 ('Customer communication')
Proof	The identification, review and interaction with customers and customer requirements
Likely Documentation	Quality Manual Quality Plans

8.2.2 Determining the requirements for products and services

This *new* clause emphasises the need for interaction between supplier and customer so that the organisation can be certain that it knows what customers' (current and future) requirements are in order that it can use this knowledge to identify the products and services it intends to offer to customers.

ISO 9001:2015 Clause 8.2.2 'Determining the requirements for products and services'	'When determining the requirements for the products and services to be offered to customers, the organisation shall ensure that:
	• the requirements for the products and services are defined, including:
	  – any applicable statutory and regulatory requirements;
	  – those considered necessary by the organisation;
	• the organisation can meet the claims for the products and services it offers'

Equivalent ISO 9001:2008 Clause(s)	7.2.1 'Determination of requirements related to the product services'
Proof	The identification, review and interaction with customers and customer requirements
Likely Documentation	Quality Manual Quality Plans – showing lines of communication with the customer contract review procedures

Before entering into a contract situation, an organisation needs to find out exactly what the customer wants regarding product specification, availability, delivery, support, etc. It also needs to confirm that it has sufficient resources to complete the contract and is capable of satisfying the customer's requirements – in full!

8.2.3 Review of the requirements for products and services

The organisation must review requirements relating to its products and services, particularly specific requirements set by the customer concerning delivery and post-delivery.

ISO 9001:2015 Clause 8.2.3 'Review of requirements for products and services'	'The organisation shall ensure that it has the ability to meet the requirements for products and services to be offered to customers and shall retain documented information, as applicable: • on the results of the review; • on any new requirements for the products and services'
Equivalent ISO 9001:2008 Clause(s)	7.2.2 'Review of requirements related to the product'
Proof	How an organisation reviews product and customer requirements to check that they can actually do the job.
Likely Documentation	• contract review procedures; • project plans showing lines of communication with the customer.

Most organisations will offer their standard products and services in a paper or (increasingly nowadays) on-line catalogue from which the customer can make a selection. These products and services will be identified against a design specification and will normally be accompanied by a picture and/or technical description. Most organisations will usually be willing to provide system-specific products to suit individual customer requirements.

8.2.4 Changes to requirements for products and services

Specialist service requirements will differ from one customer to another (and from one contract to another) and will, therefore, possibly need to be covered by an individual tender, quotation and/or contract.

ISO 9001:2015 Clause 8.2.4 'Changes to requirements for products and services'	'The organisation shall ensure that relevant documented information is amended, and that relevant persons are made aware of the changed requirements, when the requirements for products and services are changed'
Equivalent ISO 9001:2008 Clause(s)	8.2.2 'Review of requirements related to the product'
Proof	Internal audit
Likely Documentation	Quality Manual

8.3 Design and development of products and services

This is a *new* clause that *mandates* the introduction of a design and development process. It is particularly aimed at organisations that do not have established detailed requirements for their products or services, or where these have not been defined by the customer or other interested parties.

With the introduction of ISO 9001:2015, in future, *all* organisations will now be *required* to plan and control the design and development of their products and services and describe a process comprising a number of stages, each of which will be subject to controls.

8.3.1 General

Design usually means the production of something new although it can, in many circumstances, be a variation of an existing product or service. It could, therefore, be a new product or it could be a system made up of a variety of products.

ISO 9001:2015 Clause 8.3.1 General	'The organisation shall establish, implement and maintain a design and development process that is appropriate to ensure the subsequent provision of products and services'
Equivalent ISO 9001:2008 Clause(s)	None – although this clause has includes a lot of ideas previously found in 7.3 'Design and development'
Proof	The availability of a process to control design and development stages within an organisation such as: • planning; • inputs; • outputs; • review; • verification; • validation; • change control.
Likely Documentation	• processes and procedures for design and development; • design plans and development plans.

A process or design plan needs to be developed that confirms:

- what the customer needs;
- what the boundaries are (e.g. customer requirements);
- how the organisation is going to achieve it;
- how long it will take;
- who will undertake the task;
- who will check and verify the product.

Author's Hint

Although design and development is not always applicable to some small businesses (particularly if the organisation simply manufactures a product or provides a service for someone else) and it doesn't actually 'own' the design, it will probably need to develop a process in order to complete the product. In a small business, therefore, design control can be a very challenging process and methods and controls for ensuring the success of a process need to be flexible, and the process itself should be capable of being tailored to the size of any project.

8.3.2 Design and development planning

The organisation is required to plan and control the design and development of its products and services, and to describe them as a process comprising a number of stages.

ISO 9001:2015 Clause 8.3.2 'Design and development planning'	'The organisation shall establish, implement and maintain a design and development process that is appropriate to ensure the subsequent provision of products and services'
Equivalent ISO 9001:2008 Clause(s)	7.3.1 'Design and development planning'
Proof	How an organisation goes about planning and controlling the design and development of a product.
Likely Documentation	• design and development plans; • processes and procedures for design and development; • risk assessment; • job descriptions and responsibilities.

The best production methods cannot compensate for an inadequate or mediocre design! Quality cannot be an 'add-on' – it has to be designed into a product before it is created, and the only way of achieving that is through careful planning and controlled documentation throughout the design stage.

8.3.3 Design and development inputs

Following initial contract approval, details of all the relevant standards, specifications and specific customer requirements that are going to be used during production will have to be identified and steps taken to ensure that these are available.

ISO 9001:2015 Clause 8.3.3 'Design and development inputs'	'The organisation shall determine the requirements essential for the specific types of products and services to be designed and developed'.
Equivalent ISO 9001:2008 Clause(s)	7.3.2 'Design and development inputs'
Proof	How an organisation identifies the requirements to be met by a product.
Likely Documentation	Project plans detailing: • policies; • standards; • specifications and tolerances; • skill requirements; • regulatory and legal requirements; • information derived from previous (similar) • designs or developments; • environmental requirements; • health and safety aspects.

Procedures will have to be established and maintained in order to make certain that the functions of the design office are in agreement with the specified requirements. Any incomplete, ambiguous or conflicting requirements must be resolved at this stage and revisions of the specification reviewed and agreed upon by both parties.

8.3.4 Design and development controls

There are *no* new requirements (for a change!), but the organisation must ensure that design and development inputs are adequate, complete and unambiguous, and if there are any variations of 'conflicts' between design inputs, then it must ensure that these are resolved.

ISO 9001:2015 Clause 8.3.4 'Design and development controls'	'The organisation shall apply controls to the design and development process and verify that the design output meets the design and development inputs'
Equivalent ISO 9001:2008 Clause(s)	7.3.4 ('Design and development review') 7.3.5 ('Design and development verification'); and 7.3.6 ('Design and development validation')
Proof	How an organisation ensures that the product specifications are fulfilled and that the finalised product meets the original input requirements (i.e. the design is actually capable of doing the intended job).
Likely Documentation	• design process review procedures; • procedures for in-process inspection and testing; • procedures for periodic reviews; • procedures detailing how changes are made to designs and how they are approved, recorded and distributed; • management reviews and audit procedures; • final inspection and test procedures; • documented information.

8.3.5 Design and development outputs

ISO 9001:2015 Clause 8.3.5 'Design and development outputs'	'The organisation shall ensure that design and development outputs are capable of meeting the input requirements'
Equivalent ISO 9001:2008 Clause(s)	7.3.3 'Design and development outputs'
Proof	How an organisation ensures that the design output meets the design input requirements.
Likely Documentation	• drawings; • schematics; • schedules; • system specifications; • system descriptions, etc.

All documentation associated with the design output (e.g. drawings, schematics, schedules, system specifications, system descriptions, etc.) needs to:

- be produced in accordance with agreed customer requirements;
- be reviewed (by another designer who has not been associated with the initial design) to ensure that it meets the design input;
- identify all of the characteristics which are critical to the effective operation of the designed system;
- be reviewed and approved by the customer prior to use.

8.3.6 Design and development changes

ISO 9001:2015 Clause 8.3.6 'Design and development changes'	'The organisation shall identify, review and control changes made during, or subsequent to, the design and development of products and services, to the extent necessary to ensure that there is no adverse impact on conformity to requirements'
Equivalent ISO 9001:2008 Clause(s)	7.3.7 'Control of design and development changes'
Proof	How changes to a design are identified, evaluated, recorded and approved.
Likely Documentation	• change control procedures; • design process review procedures; • management reviews and audit procedures.

Throughout the design and development phase, there are likely to be a number of changes, alterations, modifications and improvements made to the design of the product and its development processes. It is essential that:

- these are identified, documented and controlled;
- the effect of the changes on constituent parts and delivered products is evaluated;
- the changes are verified, validated and approved before implementation.

8.4 Control of externally provided processes, products and services

 Supplier selection and verification can be an absolute minefield for small businesses and the chosen process needs careful consideration – particularly with respect to reliability, availability and mutual respect.

ISO 9001:2015 Clause 8.4 'Control of externally provided processes, products and services' – General	'The organisation shall ensure that externally provided processes, products and services conform to requirements and that they do not adversely affect the organisation's ability to consistently deliver conforming products and services to its customers'
Equivalent ISO 9001:2008 Clause(s)	7.4.1 'Purchasing process'
Proof	How an organisation controls the purchase of materials, products and services from suppliers and third parties.
Likely Documentation	• documented procedures for the evaluation of suppliers; • documented procedures for the evaluation of a purchased product or service.

8.4.1 General

ISO 9001:2015 Clause 8.4.1 'General'	'The organisation shall ensure that externally provided processes, products and services conform to requirements'
Equivalent ISO 9001:2008 Clause(s)	7.4.1 'Purchasing process'
Proof	The controls that an organisation has in place to ensure purchased products and services are of an acceptable standard.
Likely Documentation	<ul><li>approved list of suppliers;</li><li>supplier evaluations;</li><li>purchasing procedures;</li><li>purchase orders.</li></ul>

Purchasing processes and procedures

The organisation is responsible for producing purchasing processes and procedures that include:

- identification of requirements;
- selection of suppliers;
- quotations and tenders;
- purchase price;
- order forms;
- verification of purchased products;
- nonconforming purchased products;
- contract administration and associated purchase documentation;
- supplier control and development;
- risk assessment.

8.4.2 Type and extent of control

 The organisation shall evaluate and select suppliers based on their ability to supply products in accordance with the organisation's requirements.

ISO 9001:2015 Clause 8.4.2 'Type and extent of control'	'The organisation shall ensure that externally provided processes, products and services do not adversely affect the organisation's ability to consistently deliver conforming products and services to its customers'
Equivalent ISO 9001:2008 Clause(s)	7.4.1 ('Purchasing process') and 7.4.3 ('Verification of purchased product')
Proof	The controls that an organisation has in place to ensure that products and services provided by suppliers meet their original requirements.
Likely Documentation	• approved list of suppliers; • supplier evaluations; • purchasing procedures; • purchase orders; • stock control procedures.

Having identified its suppliers (usually selected from previous experience, past history, test results on similar projects or published experience from other users), an organisation should establish a system by which the supplier/subcontractor is clearly advised exactly what is required, and by what date. This is often achieved by use of a purchase order system.

8.4.3 Information for external providers

ISO 9001:2015 Clause 8.4.3 'Information for external providers'	'The organisation shall ensure the adequacy of requirements prior to their communication to the external provider'
Equivalent ISO 9001:2008 Clause(s)	7.4.1 ('Purchasing processes'); 7.4.2 ('Purchasing information'); and 7.4.3 ('Verification of purchased product')
Proof	The details that need to be provided by an organisation when placing an order with a supplier.
Likely Documentation	• approved list of suppliers; • supplier evaluations; • purchasing procedures; • purchase orders; • stock control procedures.

A process should be established to ensure that purchasing documents contain sufficient details about:

- the product to be purchased;
- the necessary approval and qualification requirements (i.e. procedures, processes, etc.) for product, equipment and personnel;
- the QMS requirements;
- agreement on Quality Assurance (QA) – whether the prime contractor can rely completely on the subcontractor's QA scheme or whether some (or all) of the product will have to be tested by the prime contractor or via a third party;
- agreement on verification methods by the purchaser at source or on delivery;
- whether this should be by sample or on a 100% basis;
- whether this inspection should be at the prime contractor's or the sub-contractor's premises;
- settlement of quality disputes – who, how, when and where.

8.5 Production and service provision

This clause requires the organisation to control the way in which it provides its products and services and what activities need to be performed to produce them.

8.5.1 Control of production and service provision

ISO 9001:2015 Clause 8.5.1 'Production and service provision'	'The organisation shall implement production and service provision under controlled conditions'
Equivalent ISO 9001:2008 Clause(s)	7.5.1 ('Control of production and service provision') and 7.5.2 ('Validation of processes from production and service provision')
Proof	The availability of all relevant information concerning control production and service operations. How an organisation identifies processes which cannot be verified by subsequent monitoring/ testing/inspection (including the validation of these processes to demonstrate their effectiveness).
Likely Documentation	• procedures; • project plans; • resources; • processes; • Quality Procedures; • Work Instructions.

The organisation should identify the requirements for product realisation and ensure that it has:

- the ability to comply with contractual requirements;
- the ability to train and have available competent people;
- a viable system for communication;
- a process for problem prevention.

8.5.2 Identification and traceability

ISO 9001:2015 requires organisations to maintain documented procedures for identifying products (hardware, software, documents and/or data) throughout all stages of production, delivery, receipt and installation.

ISO 9001:2015 Clause 8.5.2 'Identification and traceability'	'The organisation shall use suitable means to identify outputs when it is necessary to ensure the conformity of products and services'
Equivalent ISO 9001:2008 Clause(s)	7.5.3 'Identification and traceability'
Proof	How the status of products and services are identified during all stages of its production/delivery
Likely Documentation	Documented: • processes; • Quality Procedures; • Work Instructions.

This process should be documented and reviewed for its continued applicability on a regular basis.

8.5.3 Property belonging to customers or external providers

Products received from customers need to be visually inspected at the receipt stage and any undeclared nonconformance immediately reported to the customer.

ISO 9001:2015 Clause 8.5.3 'Property belonging to customers or external providers'	'The organisation shall exercise care with property belonging to customers or external providers while it is under the organisation's control or being used by the organisation'
Equivalent ISO 9001:2008 Clause(s)	7.5.4 'Customer property'
Proof	How an organisation looks after property that has been provided by a customer.
Likely Documentation	A documented procedure for the control of customer property.

The organisation should ensure that all property belonging to the customer is protected and that care is taken to ensure that it is well maintained, used in accordance with the supplier's instructions and safeguarded at all times.

 Author's Hint
'Customer property' can sometimes include 'intellectual property'!

8.5.4 Preservation

A manufacturer's/supplier's part number or description label should identify any material or equipment that cannot be obviously identified.

ISO 9001:2015 Clause 8.5.4 'Preservation'	**'The organisation shall preserve the outputs during production and service provision, to the extent necessary to ensure conformity to requirements'**
Equivalent ISO 9001:2008 Clause(s)	7.5.5 'Preservation of product'
Proof	How an organisation looks after its products.
Likely Documentation	• product approval procedures; • procedures which ensure the safety and protection of products.

This identification can be on the packaging or on the item itself, and should remain in place for as long as possible provided it does not hamper effective use of the item. If items have a serial number then this number should also be recorded.

8.5.5 Post-delivery activities

Following the delivery of a product or service, there might be some additional post-delivery activities that need to be considered such as actions under warranty provisions, contractual obligations (e.g. maintenance services) and supplementary services such as recycling or final disposal.

ISO 9001:2015 Clause 8.5.5 'Post-delivery activities'	**'The organisation shall meet requirements for post-delivery activities associated with the products and services'.**
Equivalent ISO 9001:2008 Clause(s)	7.5.1 'Control of production and service provision'
Proof	How an organisation determines and implements customer, product and regulatory requirements.
Likely Documentation	Quality Manual Quality Plan.

8.5.6 Control of changes

ISO 9001:2015 Clause 8.5.6 'Control of changes'	'The organisation shall review and control changes for production or service provision, to the extent necessary to ensure continuing conformity with requirements'
Equivalent ISO 9001:2008 Clause(s)	7.3.7 'Control of design and development changes'
Proof	A documented procedure showing how an organisation is capable of continuously producing quality products and/or services.
Likely Documentation	Documented information such as a change control procedure.

Any changes, alterations, modifications and improvements that have been made to the design and use of the product or service need to be properly identified, documented and controlled.

8.6 Release of products and services

The organisation needs to establish, specify and plan their measurement requirements (including acceptance criteria) prior to the release of its products and services.

ISO 9001:2015 Clause 8.6 'Release of products and services'	'The organisation shall implement planned arrangements, at appropriate stages, to verify that the product and service requirements have been met'
Equivalent ISO 9001:2008 Clause(s)	8.2.4 ('Monitoring and measurement of processes'); and 7.4.3 ('Verification of purchased product')
Proof	How an organisation ensures that product characteristics meet the customer's specified requirements.
Likely Documentation	• audit schedules; • audit plans, check sheets and records; • approval procedures for product acceptance; • processes for failure cost analysis, conformity, nonconformity, life cycle approach and self-assessment; • compliance with environmental and safety policies, laws, regulations and standards; • procedures for testing and monitoring processes; • performance and product measurement procedures; • supplier approval procedures.

Typical examples of product measurement records include:

- inspection and test reports;
- material release notices;
- certificates as required;
- electronic data.

8.7 Control of nonconforming outputs

The organisation needs to develop sufficient controls to identify any process, product or service that does not meet specified requirements and is *not* delivered to the customer or used unintentionally.

ISO 9001:2015 Clause 8.7 'Control of nonconforming process outputs' (Part 8.7.1)	'The organisation shall ensure that outputs that do not conform to their requirements are identified and controlled to prevent their unintended use or delivery'
Equivalent ISO 9001:2008 Clause(s)	8.3 'Control of nonconforming products and services'
Proof	The methods used to prevent the use or delivery of nonconforming products and services as well as their storage and disposal.
Likely Documentation	• documented procedure to identify and control the use and delivery of nonconforming products and services; • approval procedures; • quarantine procedures; • change control procedure; • corrective and preventive action procedure.

Controls need to be established and implemented to ensure that these 'nonconforming' process outputs, products or services are not delivered to the customer or used unintentionally.

ISO 9001:2015 Clause 8.7 'Control of nonconforming process outputs' (Part 8.7.2)	'The organisation shall retain documented information that: • describes the nonconformity; • describes the actions taken; • describes any concessions obtained; • identifies the authority deciding the action in respect of the nonconformity'
Equivalent ISO 9001:2008 Clause(s)	8.3 'Control of nonconforming products and services'
Proof	The methods used to prevent the use or delivery of nonconforming products, as well as their storage and disposal.
Likely Documentation	• documented procedure to identify and control the use and delivery of nonconforming products; • approval procedures; • quarantine procedures; • change control procedure; • corrective and preventive action procedure.

9 PERFORMANCE EVALUATION

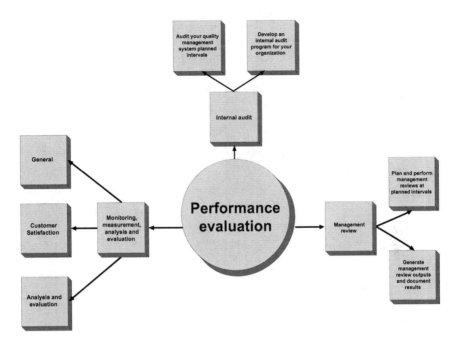

FIG. A.7 Clause 9: performance evaluation

9.1 Monitoring, measurement, analysis and evaluation

This clause (and its sub-clauses) shows how an organisation:

- determines what, how and when things are to be monitored, measured, analysed and evaluated;
- conducts its internal audits so as to ensure that its management system conforms to the requirements of the organisation as well as the selected management standard;
- ensures that its management system is successfully implemented and maintained;
- organises and conducts its management reviews to see whether they are, and can remain, suitable, adequate and effective.

9.1.1 General

Basically this clause is really all about risk assessment where, having initially determined when, how and what the organisation needs to monitor and measure, it can then make a decision how best to carry out these activities in order to improve the quality performance and effectiveness of its QMS.

ISO 9001:2015 Clause 9.1.1 'General'	'The organisation shall determine: • **what needs to be monitored and measured;** • **the methods for monitoring, measurement, analysis and evaluation needed to ensure valid results;** • **when the monitoring and measuring shall be performed;** • **when the results from monitoring and measurement shall be analysed and evaluated'**
Equivalent ISO 9001:2008 Clause(s)	8.1 ('General') and 8.2 ('Monitoring and measurement')
Proof	Documented procedures to ensure conformity, improvement, analysis of customer satisfaction and the control of products, services and processes
Likely Documentation	Procedures for: • products and/or services conformity; • products and/or services improvement; • statistical process review; • inspection and measurement; • risk analysis.

9.1.2 Customer satisfaction

The principal change here is that the organisation can no longer make its own decision (perception) as to whether it has satisfied its customers' requirements; it now needs to find out exactly what the customer thinks of the organisation, its products and services.

ISO 9001:2015 Clause 9.1.2 'Customer satisfaction'	**'The organisation shall monitor customers' perceptions of the degree to which their needs and expectations have been fulfilled.** **The organisation shall determine the methods for obtaining, monitoring and reviewing this information'**
Equivalent ISO 9001:2008 Clause(s)	8.2.1 'Customer satisfaction'
Proof	The processes used to establish whether a customer is satisfied with a product.
Likely Documentation	Procedures for: • customer feedback; • change control; • customer complaints.

The organisation must establish a process (or processes) to gather, analyse and make effective use of all customer-related information as one of the measurements of its QMS performance. This information can come from multiple sources such as:

- customer requirements and contract information;
- feedback from the delivery of a product;
- market needs;
- service delivery data;
- information relating to competition.

9.1.3 Analysis and evaluation

Customers may often require confirmation that the organisation is capable of continuing to produce a quality product, service or process.

ISO 9001:2015 Clause 9.1.3 'Analysis and evaluation'	'The organisation shall analyse and evaluate appropriate data and information arising from monitoring and measurement'
Equivalent ISO 9001:2008 Clause(s)	8.4 'Analysis of data'
Proof	The methods used to review data that will determine the effectiveness of the QMS.
Likely Documentation	Data or statistics produced as a result of audits, customer satisfaction surveys, complaints, nonconformances, supplier evaluations, etc.

Author's Hint

One of the methods frequently used to provide this sort of confirmation is statistical analysis.

9.2 Internal audit

Although organisations no longer have a mandatory requirement to provide a Quality Procedure for internal audits they are, nevertheless, still required to carry out internal audits at planned intervals in order to determine whether their QMS conforms to both the organisation's own requirements and the requirements of ISO 9001:2015.

ISO 9001:2015 Clause 9.2 'Internal audit' (Part 9.2.1)	**'The organisation shall conduct internal audits at planned intervals to provide information on whether the Quality Management System conforms to:** • **the organisation's own requirements for its QMS;** • **the requirements of this international standard'**
Equivalent ISO 9001:2008 Clause(s)	8.2.2 'Internal audit'
Proof	The in-house checks made to determine if the QMS is: • functioning properly; • continuing to comply with the requirements of ISO 9001:2015.
Likely Documentation	• audit procedures; • audit schedules; • audit plans, check sheets and records.

The method for conducting these internal audits will be as shown below in the second sub-clause to 9.2.

ISO 9001:2015 Clause 9.2 'Internal audit' (Part 9.2.2)	'The organisation shall plan, establish, implement and maintain an audit programme including the frequency, methods, responsibilities, planning requirements and reporting, which shall take into consideration the importance of the processes concerned, changes affecting the organisation and the results of previous audits'
Equivalent ISO 9001:2008 Clause(s)	8.2.2 'Internal audit'
Proof	Documented information covering the organisation's policy concerning internal audits
Likely Documentation	Quality Manual

The organisation will need to establish an internal audit process to assess the strengths and weaknesses of their QMS, to identify potential danger spots, eliminate wastage and verify that corrective action has been successfully achieved.

9.3 Management review

Top Management still need to conduct regular reviews of their QMS at planned intervals, paying special attention to the new items (added by ISO 9001:2015) relating to context, risks and opportunities that have been included.

9.3.1 General

ISO 9001:2015 Clause 9.3.1 'General'	'Top Management shall review the organisation's Quality Management System, at planned intervals, to ensure its continuing suitability, adequacy, effectiveness and alignment with the strategic direction of the organisation'
Equivalent ISO 9001:2008 Clause(s)	5.6 'Management review'
Proof	How Top Management reviews the QMS to ensure its continued suitability, adequacy and effectiveness, in the context of an organisation's strategic planning cycle.
Likely Documentation	• management review; • QMS audit procedures.

Management need to establish a process for periodically reviewing the organisation's QMS to ensure that it continues to meet the requirements of ISO 9001:2015, agrees with the organisation's policies and objectives and continues to provide customer satisfaction. Current performance, client feedback and opportunities for improvement also need to be evaluated, and possible alterations have to be made to the relevant quality documentation analysed.

9.3.2 Management review inputs

For completeness, inputs to management reviews should include everything concerned with the performance, conformance and improvement of the product.

ISO 9001:2015 Clause 9.3.2 'Management review inputs'	'The management review shall be planned and carried out taking into consideration: • actions from previous management reviews; • changes in external and internal issues relevant to the QMS; • customer satisfaction and feedback from relevant interested parties; • process performance and conformity of products and services; • nonconformities and corrective actions; • monitoring and measurement results; • audit results; • performance of external providers; • adequacy of resources; • effectiveness of actions taken to address risks; • opportunities for improvement'
Equivalent ISO 9001:2008 Clause(s)	5.6 'Management review'
Proof	How a Top Management review of the QMS is completed.
Likely Documentation	Results of audits, customer feedback, analysis of product conformance with process and procedural rules, corrective and preventive action reports and supplier performance records.

Inputs would include, but are not restricted to:

- evaluation of product conformance;
- results from previous internal, customer and third party audits;
- study of process performance;
- supplier performance;
- the current status of corrective and preventive actions;
- the results of self-assessment of the organisation;
- the review of customer feedback.

9.3.3 Management review outputs

The aim of completing management reviews is to provide a continuing record of the organisation's capability to produce quality products that meet the quality objectives, policies and requirements contained in their QMS.

ISO 9001:2015 Clause 9.3.3 'Management review outputs'	'The outputs of the management review shall include decisions and actions related to: • opportunities for improvement; • any need for changes to the Quality Management System; • resource needs'
Equivalent ISO 9001:2008 Clause(s)	5.6 'Management review'
Proof	How the results of management reviews of the QMS are documented.
Likely Documentation	Minutes of the Meetings where the overall running of the company is discussed.

Review output should be centred on:

- improved product and process performance;
- conformation of resource requirements and organisational structure;
- meeting market needs;
- risk management;
- change control;
- continued compliance with relevant statutory and regulatory requirements.

10 IMPROVEMENT

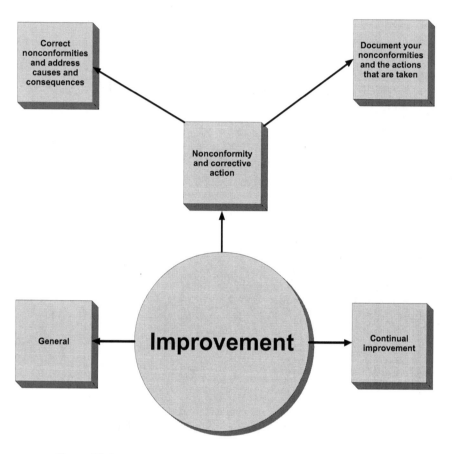

FIG. A.8 Clause 10: improvement

This is a *new* clause which requires organisations to proactively look for opportunities that will enable them to meet customer requirements and enhance customer satisfaction.

10.1 General

It is a well-known fact that in an ever-changing business world, not everything will always go according to plan, and so Clause 10 looks at ways to address non-conformities and corrective action, as well as suggesting strategies for improvement on a continual basis.

ISO 9001:2015 Clause 10.1 'General'	'The organisation shall determine and select opportunities for improvement and implement any necessary actions to meet customer requirements and enhance customer satisfaction'
Equivalent ISO 9001:2008 Clause(s)	8.5.1 'Continual improvement'
Proof	The methods used to prevent the use or delivery of nonconforming products, as well as their storage and disposal. How an organisation controls corrective and preventive actions and how it ensures the continual improvement of its product.
Likely Documentation	• documented procedure to identify and control the use and delivery of nonconforming products; • approval procedures; • quarantine procedures; • change control procedure; • corrective and preventive action.

10.2 Nonconformity and corrective action

To ensure that nonconforming or hazardous products or services are not delivered by mistake to a customer, the organisation must establish and maintain procedures for identifying products and services (from drawings, specifications or other documents) during all stages of production, delivery and installation.

10.2.1

ISO 9001:2015 Clause 10.2 'Nonconformity and corrective action' (Part 10.2.1)	'When a nonconformity, including any arising from complaints, occurs the organisation shall take action to control and correct it'
Equivalent ISO 9001:2008 Clause(s)	8.3 ('Control of nonconforming product'); and 8.5.2 ('Corrective action')
Proof	The methods used to prevent the use or delivery of nonconforming products and services, as well as their storage and disposal. What an organisation does to identify and put right nonconformities.
Likely Documentation	• documented procedure to identify and control the use and delivery of nonconforming products and services; • approval procedures; • quarantine procedures; • change control procedure; • corrective and preventive action; • process for eliminating causes of nonconformity; • documented complaints; • complaints procedure; • Staff suggestions scheme.

10.2.2

ISO 9001:2015 Clause 10.2 'Nonconformity and corrective action' (Part 10.2.2)	'The organisation shall retain documented information as evidence of: • the nature of the nonconformities; • any subsequent actions taken; • results of any corrective action'
Equivalent ISO 9001:2008 Clause(s)	8.3 ('Control of nonconforming product') and 8.5.2 ('Corrective action')
Proof	A procedure to retain documented information concerning nonconforming products and services
Likely Documentation	Quality Manual

10.3 Continual improvement

The organisation should continually seek to improve its processes and procedures (rather than just waiting for a problem to come along) and have available documented procedures to identify, manage and improve them.

ISO 9001:2015 Clause 10.3 'Continual improvement'	'The organisation shall continually improve the suitability, adequacy and effectiveness of the Quality Management System'
Equivalent ISO 9001:2008 Clause(s)	8.5.1 'Continual improvement'
Proof	How an organisation goes about continually improving its QMS.
Likely Documentation	• procedures, Minutes of Meetings (where improvement to the organisation's business is discussed); • management reviews.

Abbreviations and acronyms

AFNOR	Association Francais de Normalisation (French Association for Standardisation)
AFS	Affiliate Country Programme
ANSI	American National Standards Institute
AQAP	Allied Quality Assurance Publications (NATO)
BS	British Standard, issued by BSI
BSI	British Standards Institution
CAD	Computer Aided Design
CCIR	International Radio Consultative Committee
CCITT	The International Telegraph and Telephony Consultative Committee
CD	Committee Draft
CEN	Commission European de Normalisation
CENELEC	European Committee for Electrotechnical Standardisation
COS	Corporation of Open Systems
CPS	Corporate Policy Statement
DBOM	Design, Build, and Operation & Maintenance
DEF STAN	Defence Standards (UK)
DIN	Deutsches Institut für Normung (German Institute for Standardisation)
DIS	Draft International Standard
DMAIC	Define, Measure, Analyse, Improve and Control
DOD	(American) Division of Defense
DTI	Department of Trade and Industry
EN	European Number (for European standards)
ETSI	European Telecommunications Standards Institute
EX	Explosive

FDIS	Final Draft International Standard
GMP	Good Manufacturing Practice
GQA	Government Quality Assurance
IBC	Intermediate Bulk Containers
IEC	International Electrotechnical Commission
IMS	Integrated Management Systems
IRCA	International Register of Certificated Auditors
ISMS	Information Security Management System
ISO	International Organisation for Standardisation
ITU	International Telegraph Union
LAN	Local Area Network
Mil-Std	Military Standard
MOD-UK	Ministry of Defence (UK)
MSS	Management Systems Standards
NASA	National Aeronautics and Space Administration
NATO	North Atlantic Treaty Organisation
NSO	National Standards Organisation
OHSAS	Occupational Health and Safety Management Systems
OJT	On-the-Job Training
OSI	Open Systems Interconnection
QA	Quality Assurance
QC	Quality Control
QMS	Quality Management System
QP	Quality Procedure
QuEST Forum	Quality Excellence for Suppliers of Telecommunications Leadership
PAS	Publicly Available Specifications
SME	Small and Medium-sized Enterprises
SWOT	Strengths/Weaknesses/Opportunities/Threats
TQM	Total Quality Management (BS 7850)
TS	Technical Specifications

UK	United Kingdom
UN	United Nations
WD	Working Draft
WI	Work Instruction

Reference standards for Quality Management Systems

Number	Title
ISO 9000	Quality Management Systems – Fundamentals and vocabulary
ISO 9001	Quality Management Systems – Requirements
ISO 9004	Managing for the sustained success of an organisation – A quality management approach
ISO 10001	Quality management – Customer satisfaction – Guidelines for codes of conduct for organisations
ISO 10002	Quality management – Customer satisfaction – Guidelines for complaints handling in organisations
ISO 10003	Quality management – Customer satisfaction – Guidelines for dispute resolution external to organisations
ISO 10004	Quality management – Customer satisfaction – Guidelines for monitoring and measuring
ISO 10005	Quality management – guidelines for Quality Plans
ISO 10006	Quality Management Systems – Guidelines for quality management in projects
ISO 10007	Quality Management Systems – Guidelines for configuration management
ISO 10008	Quality management – Customer satisfaction – Guidelines for business-to-consumer electronic commerce transactions
ISO 10018	Quality management – Guidelines on people involvement and competence
ISO 19011	Guidelines for the selection of Quality Management System consultants and use of their services

Complete copies of these Standards are available from ISO Member Countries in their own languages. The British versions (e.g. BS EN ISO 19011) can be obtained, by post, from Customer Services, BSI Standards, 389 Chiswick High Road, London W4 4AL.

 Note: Extracts from British Standards reproduced in this book are with kind permission of the British Standards Institute.

Glossary of terms used in Quality Management standards

Acceptable quality level: A measure of the number of failures that a production process is allowed. Usually expressed as a percentage.

Accreditation: Certification, by a duly recognised body, of facilities, capability, objectivity, competence and integrity of an agency, service or operational group or individual to provide the specific service/s or operation/s as needed.

Assemblies: Several pieces of equipment assembled by a manufacturer to constitute an integrated and functional whole.

Audit: Systematic, independent and documented process for obtaining evidence and evaluating it objectively to determine the extent to which audit criteria are fulfilled.

Audit team: One or more auditors conducting an audit, one of whom is appointed as leader.

Certification: The procedure and action by a duly authorised body of determining, verifying and attesting in writing to the qualifications of personnel, processes, procedures or items in accordance with applicable requirements.

Certification Body: An impartial body, governmental or non-governmental, possessing the necessary competence and reliability to operate a certification system, and in which the interests of all parties concerned with the functioning of the system are represented.

Chief inspector: An individual who is responsible for the manufacturer's Quality Management System.

Company: Term used primarily to refer to a business first party, the purpose of which is to supply a product or service.

Compliance: An affirmative indication or judgement that a product or service has met the requirements of the relevant specifications, contract or regulation. Also the state of meeting the requirements.

Conformance: An affirmative indication or judgement that a product or service has met the requirements of the relevant specifications, contract or regulation. Also the state of meeting the requirements.

Contract: Agreed requirements between a supplier and customer transmitted by any means.

Customer: Ultimate consumer, user, client, beneficiary or second party.

Customer satisfaction: Customer's opinion of the degree to which a transaction has met the customer's needs and expectations.

Defect: Non-fulfilment of a requirement related to an intended or specified use.

Design and development: Set of processes that transforms requirements into specified characteristics and into the specification of the product realisation process.

Distributor: An organisation that is contractually authorised by one or more manufacturers to store, repack and sell completely finished components from these manufacturers.

Document: Information and its support medium.

Documented information: Information that the organisation will be required to keep, control and maintain.

Environment: All of the external physical conditions that may influence the performance of a product or service.

Equipment: Machines, apparatus, fixed or mobile devices, control components and instrumentation thereof and detection or prevention systems which, separately or jointly, are intended for the generation, transfer, storage, measurement, control and conversion of energy for the processing of material and which are capable of causing an explosion through their own potential sources of ignition.

In-process inspection: Inspection carried out at various stages during processing.

International Organisation for Standardisation (ISO): Comprises the national standards bodies of 163 member countries whose aim is to coordinate the international harmonisation of national standards.

Item: A part, a component, equipment, sub-system or system or defined quantity of material or service that can be individually considered and separately examined or tested.

Maintenance: The combination of technical and administrative actions that are taken to retain or restore an item to a state in which it can perform its stated function.

Management: Coordinated activities to direct and control an organisation.

Management system: The establishment of policies and objectives to achieve ISO 9001:2015 objectives.

Manufacturer: An organisation which carries out or controls such stages in the manufacture of components or assemblies.

Material: A generic term covering equipment, stores, supplies and spares which form the subject of a contract.

Nonconformity: Non-fulfilment of a requirement.

Organisation: A single person or a group of people who achieve their objectives by using their own functions, responsibilities, authorities and relationships. It can be a company, corporation, enterprise, firm, partnership, charity, association or institution either privately or publicly owned. It can also be an operating unit that is part of a larger entity.

Organisational structure: Orderly arrangement of responsibilities, authorities and relationships between people.

Procedure: Describes the way to perform an activity or process.

Product: Result of a process that does not include activities that are performed at the interface between the supplier (provider) and the customer.

 Note: There are four agreed generic product categories:

- hardware (e.g. engine mechanical part);
- software (e.g. computer program);
- services (e.g. transport);
- processed materials (e.g. lubricant).

Hardware and processed materials are generally tangible products, while software or services are generally intangible.

Most products comprise elements belonging to different generic product categories. Whether the product is then called hardware, processed material, software or service depends on the dominant element.

Project: Unique process consisting of a set of coordinated and controlled activities with start and finish dates, undertaken to achieve an objective conforming to specific requirements, including the constraints of time, costs and resources.

Quality: The totality of features and characteristics of a product or service that bear upon its ability to satisfy stated or implied needs.

Quality Assurance: The assembly of all planned and systematic actions necessary to provide adequate confidence that a product, process or service will satisfy given quality requirements.

Quality characteristic: Essential characteristics of a product, process or system derived from a requirement.

Quality Control: The operational techniques and activities that are used to fulfil the requirements for quality.

Quality loop: Conceptual model of interacting activities that influence the quality of a product or service in the various stages, ranging from the identification of needs to the assessment of whether these needs have been satisfied.

Quality Manager: A person who is nominated by Top Management to be responsible for the organisation's Quality Management System (also sometimes referred to as the Chief Inspector).

Quality Management: That aspect of the overall management function that determines and implements the Quality Policy.

Quality Management System: System to establish a Quality Policy and quality objectives.

Quality Management System review: A formal evaluation by Top Management of the status and adequacy of the Quality Management System in relation to Quality Policy and new objectives resulting from changing circumstances.

Quality Manual: Document specifying the Quality Management System of an organisation and setting out its quality policies, systems and practices.

Quality Plan: Document specifying the Quality Management System elements and the resources to be applied in a specific case.

Quality Policy: The overall quality intentions and direction of an organisation as regards quality, as formally expressed by Top Management.

Quality Procedure: A description of the method by which quality system activities are managed.

Quality Process: A system which uses resources to transform inputs into outputs.

Quality system: The organisational structure, responsibilities, procedures, processes and resources for implementing Quality Management.

Requirement: Need or expectation that is stated, customarily implied or obligatory.

Review: Activity undertaken to ensure the suitability, adequacy, effectiveness and efficiency of the subject matter to achieve established objectives.

Service: Is the result of a process that includes at least one activity that is carried out at the interface between the supplier (provider) and the customer.

For example, a service might be:

- an activity performed on a customer-supplied tangible product (e.g. the repair of a car) or intangible product such as the preparation of a tax return;
- the delivery of a tangible product (e.g. in the transportation industry);
- the delivery of an intangible product (e.g. the delivery of knowledge) or the creation of ambience for the customer (e.g. in the hospitality industry).

Shall: This auxiliary verb indicates that a certain course of action is mandatory.

Should: This auxiliary verb indicates that a certain course of action is preferred but not necessarily required.

Supplier: The organisation that provides a product to the customer.
   In a contractual situation, the supplier may be called the contractor.
   The supplier may be, for example, the producer, distributor, importer, assembler or service organisation.
   The supplier may be either external or internal to the organisation.

Top Management: Person or group of people who direct and control an organisation at the highest level.

Work Instruction: A description of how a specific task is carried out.

Books by the same author

ISO 9001:2015 for Small Businesses (sixth edition)

The new edition of this top-selling Quality Management book now includes:

- Relevant examples that put the concepts and requirements of the standard into a real-life context.
- Down-to-earth explanations to help you determine what you need to work in compliance with and/or achieve certification to ISO 9001:2015.
- An example of a complete, generic, Quality Management System consisting of a Quality Manual plus a whole host of Quality Processes, Quality Procedures and Work Instructions.
- Access to a free, software copy of this generic QMS file (available from the author) to give you a starting point from which to develop your own documentation.

Routledge
ISBN-13: 978-1-138-02583-7

ISO 9001:2015 Audit Procedures (fourth edition)

Fully revised, updated and expanded, this 4th edition provides access to methods for auditing an organisation's Quality Management System against the requirements of ISO 9001:2015.

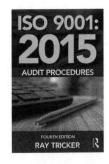

Although primarily aimed at showing how auditors from small businesses can complete management reviews and internal, external and third party quality audits, this book will prove invaluable to professional auditors.

Containing an overview of the changes made to the 2015 edition of ISO 9001 and how these will affect the way in which audits will need to be completed in future, the book also includes access to free copies of checklists, explanations and questionnaires (available from the author) that can be used for internal, external and/or third party audits of an organisation's Quality Management System.

Routledge
ISBN-13: 978-0-415-70390-1

ISO 9001:2015 in Brief (fourth edition) – this book

Now in its 4th edition, this book is particularly aimed at students, newcomers to Quality Management Systems and the busy executive, with the overall intention of providing them with a user-friendly, very simplified explanation of the history, the requirements and the benefits of the new standard.

Using this book as background material will also enable organisations (large or small) to quickly set up an ISO 9001: 2015-compliant Quality Management System for themselves – at minimal expense.

Routledge
ISBN-13: 978-1-138-02586-8

How to Convert from ISO 9001:2008 to ISO 9001:2015

The publication of ISO 9001:2015 in September 2015 signalled the start of a three-year transition period during which those organisations wishing to move to the new version of the standard were required to make changes to their existing Quality Management Systems.

'How to Convert from ISO 9001:2008 to ISO 9001:2015' provides step-by-step advice to help you through the transition and realise the benefits of ISO 9001:2015. It maps out a framework which guides you through the options and alternatives, ensuring that you have the knowledge and information you require to seamlessly make the necessary transition.

Herne European Consultancy Ltd
ISBN-13: 978-0-992-75850-9

Quality Management System for ISO 9001:2015 (fourth edition)

The Quality Management System contained in this e-Book is probably the most complete ISO 9001:2015 compliant example of a generic Quality Management System (QMS)

that can, with very little trouble, be suitably customised to suit all types of organisations – no matter whether they are manufacturers, suppliers or end users.

Consisting of a Quality Manual (supported by the four main Quality Processes, 31 Quality Procedures and 16 Work Instructions) this QMS covers every element of the standard and is guaranteed to meet (and sometimes exceed) the requirements of ISO 9001:2015.

This is an excellent resource for any small or medium sized business looking to work towards ISO certification, without having the expense of a consultant doing the work for you.

Herne European Consultancy Ltd
ISBN-13: 978-0-992-75851-6

Auditing Quality Management Systems (fourth edition)

Auditing Quality Management Systems is the result of more than four decades' experience as auditors of all major international standards used by Integrated Management Systems.

It is a comprehensive e-Book containing a series of audit checksheets and forms that are required to conduct either a simple internal audit or an external assessment of an organisation against the formal requirements of ISO 9001:2015.

Note: also includes 'Background notes for auditors'

Herne European Consultancy Ltd
ISBN-13: 978-0-992-75852-3

MDD Compliance using Quality Management Techniques

The Medical Device Directive (MDD) is difficult to understand and interpret, but this book covers the subject superlatively.

In summary, the book is a good reference for understanding the MDD's requirements and would aid companies of all sizes in adding these requirements to an existing QMS.

Butterworth Heinemann
ISBN-13: 978-0-750-64441-9

Building Regulations in Brief (eighth edition)

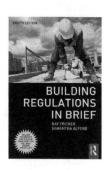

This eighth edition of the most popular and trusted guide to the building regulations is the most comprehensive revision yet. It reflects all the latest amendments to Building Regulations, Planning Permission and the Approved Documents A,B,C, H, K, P, Regulation 7, incorporating all amendments up to December 2013 (including the changes to Leaflets L1A and L2A regarding the conservation of heat and energy in new buildings which came into effect April 2014).

This new edition also contains details of the new national planning guidance system and initiatives to speed up the planning process, such as the new online planning application process. It contains an updated list of fees for planning consents and provides guidance on the changes to permitted development rights in Agricultural, Business and Residential buildings which came into force on 1 October 2013.

Giving practical information throughout on how to work with (and within) the regulations, this book enables compliance in the simplest and most cost-effective manner possible. The no-nonsense approach of Building Regulations in Brief cuts through the confusion and explains the meaning of the regulations; consequently it has become a favourite for anyone involved in the building industry, as well as those planning to have work carried out in their home.

Routledge
ISBN: 978-0-415-72171-4

Wiring Regulations in Brief (third edition)

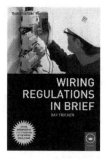

Tired of trawling through the Wiring Regs?
    Perplexed by Part P?
    Confused by cables, conductors and circuits?
    Then look no further! This handy guide provides an on-the-job reference source for electricians, designers, service engineers, inspectors, builders, students and DIY enthusiasts.

Topic-based chapters link areas of working practice – such as cables, installations, testing and inspection, special locations – with the specifics of the regulations themselves. This allows quick and easy identification of the official requirements relating to the situation in front of you.

The requirements of the regulations, and of related standards, are presented in an informal, easy-to-read style that strips away confusion.

Packed with useful hints and tips, and highlighting the most important or mandatory requirements, this book is a concise reference on all aspects of the 17th edition of IEE Wiring Regulations and Part P of the Building Regulations.

Spon Press
ISBN-13: 978-0-415-52687-6

Water Regulations in Brief

Water Regulations in Brief is a unique reference book, providing all the information needed to comply with the regulations in an easy-to-use, full-colour format.

Crucially, unlike other titles on this subject, this book doesn't just cover the Water Regulations, it also clearly shows how they link in with the Building Regulations, Water Bylaws and Wiring Regulations, providing the only available complete reference to the requirements for water fittings and water systems.

Structured in the same logical, time-saving way as the author's other bestselling '. . . in Brief' books, *Water Regulations in Brief* will be a welcome change to anyone tired of wading through complex, jargon-heavy publications in search of the information they need to get the job done.

Butterworth Heinemann
ISBN-13: 978-1-856-17628-6

Scottish Building Standards in Brief

Scottish Building Standards in Brief takes the highly successful formula of Ray Tricker's previous 'In Brief' series and applies it to the requirements of the Building (Scotland) Regulations 2004.

With the same no-nonsense and simple-to-follow guidance – but written specifically for the Scottish Building Standards – it's the ideal book for builders, architects, designers and DIY enthusiasts working in Scotland.

Routledge
ISBN-13: 978-0-750-68558-0

Quality and Standards in Electronics

A manufacturer or supplier of electronic equipment or components needs to know the precise requirements for component certification and quality conformance to meet the demands of the customer.

This book ensures that the professional is aware of all the UK, European and international necessities, knows the current status of these regulations and standards, and where to obtain them.

Newnes
ISBN-13: 978-0-750-62531-9

Environmental Requirements for Electromechanical and Electronic Equipment

This is the definitive reference containing all of the background guidance, typical ranges, details of recommended test specifications, case studies and regulations covering the environmental requirements for designers and manufacturers of electrical and electromechanical equipment worldwide.

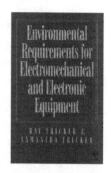

Newnes
ISBN-13: 978-0-750-63902-6

CE Conformity Marking

CE Conformity Marking can be regarded as a product's trade passport for Europe. It is a mandatory European marking for certain product groups to indicate conformity with the essential health and safety requirements set out in the European Directive.

This book contains essential information for any manufacturer or distributor wishing to trade in the European Union.

Practical and easy to understand.

Butterworth Heinemann
ISBN-13: 978-0-750-64813-4

And for those who would like to relax with some cooking recipes – based on cider and apples!

The Cyder Book

A unique combination of an historical overview of cider making through the ages, the cider-making process and a collection of recipes using cider and cider apples.

Herne European Consultancy, Ltd
ISBN-13: 978-0-954-86476-7